Dynamic Climatology

T0207446

Environmental Systems

Series editor: Antony R. Orme, *University of California, Los Angeles*

Environmental Systems is a new series of accessible, timely texts for advanced undergraduate courses. Each title will reflect current scholarship in a particular field of Environmental Systems. Written from an interdisciplinary or holistic, scientific perspective they are integrated texts in which particular components of the environment will be treated as part of the environment as a whole.

Published

Dynamic Climatology
John N. Rayner

Forthcoming titles

Glacial Systems
John T. Andrews

Weathering and Soil Systems
John C. Dixon

Deserts and Desert Environments
Julie Laity and Amalie Jo Orme

Mountain Landscapes
George P. Malanson and David R. Butler

Dynamic Climatology

Basis in Mathematics and Physics

John N. Rayner
The Ohio State University

Copyright © John N. Rayner 2001

The right of John N. Rayner to be identified as author of this work has been asserted in accordance
with the Copyright, Designs and Patents Act 1988.

First published 2001

2 4 6 8 10 9 7 5 3 1

Blackwell Publishers Ltd
108 Cowley Road
Oxford OX4 1JF
UK

Blackwell Publishers Inc.
350 Main Street
Malden, Massachusetts 02148
USA

All rights reserved. Except for the quotation of short passages for the purposes of criticism and
review, no part of this publication may be reproduced, stored in a retrieval system, or transmitted, in
any form or by any means, electronic, mechanical, photocopying, recording or otherwise, without the
prior permission of the publisher.

Except in the United States of America, this book is sold subject to the condition that it shall not, by
way of trade or otherwise, be lent, resold, hired out, or otherwise circulated without the publisher's
prior consent in any form or binding or cover other than that in which it is published and without a
similar condition including this condition being imposed on the subsequent purchaser.

British Library Cataloging in Publication Data

A CIP catalogue record for this book is available from the
British Library.

Library of Congress Cataloging-in-Publication Data

Rayner, John N.
    Dynamic climatology: basis in mathematics and physics / John N. Rayner.
        p. cm.—(Environmental systems)
    Includes bibliographical references and index.
    ISBN 1-57718-015-1 (alk. paper)—ISBN 1-57718-016-X (pb: alk. paper)
        1. Climatology. I. Title. II. Series.

QC863 .R35 2000
551.6—dc21

00-037833

Typeset in 10½ on 12 pt Times Roman
by Newgen Imaging Systems (P) Ltd, Chennai, India

www.biddles.co.uk

This book is printed on acid-free paper.

to Valerie, Caroline, Alexander, and Jonathan

Contents

Figures

Tables

Preface

People have always been fascinated by the weather, especially extreme events: ones producing disasters such as flash floods, tornadoes, and hurricanes. Climate has been a topic of less topical concern until recently when global warming and its possible effects through El Niño, rise of sea level from glacial melting, and declining agricultural production became an issue in the popular media. Whereas weather forecasting has made enormous advances in the last century, climate forecasting is still in its infancy. Even those who are professional climatologists cannot fully explain current climates, and their attempts at prediction are primitive. As a result, the general public and politicians are often confused by the experts who sometimes seem to say opposite things about future climates.

This book is written to provide a fundamental background for those interested in understanding climate. That is, it provides the historical introduction to the scientific concepts used in climatology. It is designed for two groups: those who have little relevant scientific training but want to build their background knowledge of the necessary concepts and associated terminologies without taking several courses and/or reading the relevant textbooks; and those who want a quick review of the underlying principles. For the former group the first chapters summarize ideas from several courses in mathematics, statistics and physics. On the one hand the material is very elementary and could have been covered in high school. On the other hand, because it is "classical" and very selective, it is frequently buried in several suites of courses that are often difficult to schedule in an already full curriculum. The later chapters cover the ideas more directly relevant to climate. In the latter group are students specializing in climatology who ask, "What am I supposed to know?" The answer varies depending upon the situation but these pages are designed to bring the basic material together in one place. Too often I have found that students in masters' and doctoral general examinations have not put their studies in perspective. They have usually successfully completed sequences of advanced prerequisite courses as well as the typical atmospheric science courses on radiation, dynamics, thermodynamics, and synoptics, etc., but they are invariably inept at answering relatively simple questions especially where they must combine disparate concepts. Hence this book might be considered as a first study guide. The succeeding ones are the textbooks used and notes written in the courses just mentioned.

As described within, the understanding of climate relies upon the computer. Indeed, much of my own research over the years since 1959 has depended upon the computer, initially IBM mainframes and then Cray supercomputers. Since the early 1990s more and more work has been possible on the PC. For me this has depended upon the development 'Linux,' the free unix operating system for the PC described, for example, by Welsh and Kaufman (1995). For the version I have used since 1994 I thank Linus Torvalds for the kernel, Patrick Volkerding et al. (1996) for the Slackware distribution, and Richard Stallman and the many other people in the Free Software Foundation, who developed much of the software I use, including gnu (Robbins, 1994). This book was written using Emacs (Stallman, 1997) under this operating system using the LaTeX typesetting program Lamport (1994); Goossens et al. (1994); Knuth (1986). The diagrams were produced in Postscript using Gnu Fortran, Pgplot, Xfig, and NCARgraphics.

Throughout, reference might have been made to numerous valuable web sites for data, model sequences, course notes and other excellent documentation. Unfortunately, such sources tend to be ephemeral. Also, they often tend to change their character over time as different individuals take responsibility for their administration. Hence the bibliography contains pointers to printed matter alone.

For my fundamental training I have many individuals to thank especially at schools and Norwich City College. There was Miss Isabel Montford, who had been English governess to the Kaiser's children. In order to teach me German and French she first had to teach me English. For appreciation for history and literature I thank K. Johnson of Knapton School and Messrs. Davenport, D. Duce and A. O. Browning of Norwich City College. For mathematics there were Major T. H. Russell of Cliff Holm School and S. W. Bell and H. Matley of Norwich City College. My early work in climatology was guided by E. T. Stringer at Birmingham and, due to a British Council Grant and an Arctic Carnegie Scholarship, by S. Orvig and F. K. Hare at McGill.

For this book I thank the reviewers, especially James E. Burt, who pointed out errors and made very helpful suggestions. Of course, any remaining inaccuracies and inconsistencies are mine. I am also pleased to acknowledge the invitation from a Birmingham era associate and friend, the editor of this series, Antony R. Orme, to put my ideas into print.

John N. Rayner
Dublin, OH

Chapter 1

The Field of Dynamic Climatology

Dynamic climatology is the science that endeavors to understand climates, what they are: past, present, and future; and to explain them in terms of the factors that control them. As such it is an old enterprise occupying individuals in many cultures. We all experience weather and its combined effect which we call climate. Today it is a science using the latest in satellite and electronic hardware.

The earliest written statements about atmospheric phenomena come from the eastern Mediterranean, from Egypt and Babylonia, and are at least 3,000 years old. Fairly continuous records of atmospheric events are available in China for the last 1,000 years (Liu and Shen, 1998). The word *climate*, as with many English words, may be traced back to the Greek "$\kappa\lambda\iota\mu\alpha$" or "$\kappa\lambda\iota\mu\alpha\tau$-" referring to slope as on the side of a hill but also as in latitude in relation to solar radiation. The resulting climatic classification was consequently based upon lines of latitude.

The word *dynamic* also derives from Greek where it appears as "$\delta\acute{\upsilon}\nu\alpha\mu\iota\kappa\acute{o}\varsigma$" meaning "powerful". The "$\acute{\upsilon}$"(y) is short as still sometimes heard in "dynasty" so it was pronounced "dinamic" at least until the middle of the nineteenth century. Today this adjective may be interpreted as relating to physical force or energy, i.e. of or pertaining to force producing motion: often in contrast to static.

The two words together *dynamic climatology* were introduced explicitly by Tor Bergeron in the title of his presentation, "Richtlinen einer dynamischen Klimatologie," (Guidelines for a dynamic climatology) to the German Meteorological Society in Dresden, on October 7, 1929 (Bergeron, 1930). In this paper, Bergeron called on the meteorological community to go beyond the purely descriptive aspects of climatology and put more effort into explaining climate. No English translation exists but a summary was published by Willett in 1931 in the *Monthly Weather Review* in which he stated Bergeron's main thesis as follows:

> Hitherto climatology has been essentially the systematic compiling of statistics on the individual meteorological elements, without much organized attempt to get at the underlying dynamic or thermodynamic phenomena in their entirety. We have complete charts of the distribution of atmospheric pressure, rainfall, temperature, wind, cloudiness, etc., but usually very little idea of just what sort of atmospheric activity, in toto, is behind these distributions. Willett (1931)

Willett further elaborated on Bergeron's position that the current "climatology offers no unifying picture of the prime thermodynamic forces controlling the climate. This is what a dynamic climatology should do."

Bergeron (1891–1977), born in England but brought up in Sweden, had worked in Bergen, Norway, with the Bjerkneses, Solberg and Rossby from 1919 into the 1920s when they all were researching synoptic weather forecasting. Bergeron is credited with having discovered the occluded front and with the design of the frontal symbols as well as the ice nucleation process which carries his name. He completed his doctoral dissertation at the University of Oslo in 1928 based upon the analysis of air masses. Hence his illustrations of what he considered to be "dynamic climatology" in his 1929 paper were taken mainly from that body of knowledge. There subsequently followed a large number of papers, both in German and in English, applying air mass ideas (Rayner et al., 1991). The use of such concepts in the United States, accompanied by photographs of many of the Scandinavian originators including Bergeron, was discussed by Namias (1983).

Of course, many people had attempted to explain climate well before Bergeron's time. For example, in 1784 Benjamin Franklin suggested that the introduction of volcanic dust into the atmosphere could lead to climatic changes:

> During several of the summer months of the year 1783, when the effects of the sun's rays to heat the earth in these northern regions should have been greatest, there existed a constant fog over all Europe, and great part of North America . . . Hence perhaps the winter of 1783–4 was more severe than any that happened for many years . . . whether it was the vast quantity of smoke, long continuing to issue during the summer from Hecla, in Iceland, and that other volcano which arose out of the sea near that island, which smoke might be spread by the winds over the northern part of the world, is yet uncertain. (Abbe, 1906)

The idea is still debated (Harington, 1992).

The development of geology in the early nineteenth century and the recognition that the climates of the earth had changed led to considerable speculation with regard to causes. In 1837 Louis Agassiz proposed that warming following a glacial period might have been due to mountain building (Agassiz, 1967). In 1864 Croll (1864) reviewed several possible factors, including astronomical ones, which have recently been given new life by the resurrected Milankovitch theory (Imbrie and Imbrie, 1980). In the same year Frankland (1861) introduced the argument that the latent heat inertia of oceans had a controlling effect on climatic change.

By 1873 in his *Principles of Geology* Lyell was able to review a number of theories. He made the case for the effect of orography:

Having shown the reader that there have been endless changes in the form of the earth's crust in geological times, whereby the position as well as the height and depth of the land and sea has been made to vary incessantly, and that on these geographical conditions the temperature of the atmosphere and of the ocean in any given region and at any given period must mainly depend, I shall next proceed to speculate on the changes ... (Lyell, 1873)

He also anticipated the idea of teleconnections, the effect of elements in one region on the elements in a more distant region: "in speculating on a change of climate due to altered geographical conditions, it is too often assumed that the alteration must have taken place in the immediate region where the temperature has been modified."

Although the emphasis here is on climate, much space is devoted to discussions of the underlying physics of weather. This is because weather and climate are part of an integrated continuum, weather being the instantaneous state of the atmosphere and climate being a longer term summary. Climate was defined by Durst (1951) as "the synthesis of weather," by Landsberg (1960) as "the collective state of the earth's atmosphere for a given place within a specified interval of time," and by Lorenz (1975) as "a set of statistics of an ensemble of many states of the atmosphere." Indeed, Bergeron claimed that "no hard and fast line can be drawn between weather and climate, either in definition or in controlling factors" (Willett, 1931). As noted by Lorenz, one link between weather and climate is statistics, so chapter 3 is devoted to a review of some of those techniques.

To understand climate means to be able to explain the relative importance of the various factors affecting the climate of the earth as a whole and of the various regional and subregional climates. Much of the current research into global change is aimed at the processes by which increased greenhouse gases really change global and regional temperatures. On the surface it would appear to be a simple cause and effect relationship (more absorption of long-wave radiation leads to a higher equilibrium temperature). Especially because of feedback processes, this is not necessarily the case. Very little research so far has gone into more complex issues such as the influence of the relief and land and sea distributions on climate. We know that the major mountain chains affect both temperature and moisture patterns around the globe but the extent of their influence in each region and their relative importance with respect to the other factors is unknown. Certainly climates of past geological eras were quite different because the distributions of land and sea have changed significantly.

Scientific research advances as a result of a mix of observation and the development of theoretical constructs that are eventually subject to testing. We may recognize three methodologies: (1) laboratory experiment and modeling; (2) analytical solutions of the governing equations; and (3) numerical solution of those equations, i.e. simulation.

Experiments where the various influencing factors may be controlled, requires that the system be physically moved into the laboratory. This is generally impossible for atmospheric systems although simple relationships such as between pressure, temperature, and density may be studied there. Even chemical reaction investigations are limited because of the introduction of artificial walls (the free atmosphere has none). Modeling also has a problem with size, although the manipulation of various non-dimensional parameters has produced valuable research into fluid flow. At regional and global scales the effect of earth curvature introduces serious complications.

The relevant equations relating the atmospheric variables together have been known for 150 years but they are non-linear and there are currently no general analytical solutions available. Consequently, of these methods, the third, numerical solution of these equations, is the only viable approach for the atmosphere.

Not until simulation models became sufficiently sophisticated could any real progress be made. With the development of numerical forecasting in the 1950s and 1960s it was realized that the climatology was a critical ingredient both for initialization and for gauging the accuracy of the results. Also, once the meteorological community embraced global warming as a significant research topic, the understanding of climate became a more respectable pursuit. Even so, at the dawn of the twenty-first century we are still a long way from fully understanding climate.

The reasons are several:

1 The atmosphere is global. This means that it is possible for an atmospheric event in one location eventually to affect every other location. This was clearly demonstrated in the 1950s when radioactive materials from USSR tests in the Arctic and US/British tests in low latitudes were subsequently observed in every region surveyed throughout the world (Davidson et al.,

1966). The global nature of atmospheric interactions was popularized by Michael Crichton (1990) in the movie version of *Jurassic Park* with the statement, "the butterfly that flaps its wings in Beijing and in Central Park [New York] you get rain instead of sunshine." The context was a discussion of the limits of predictability and chaos theory which was originated by Edward Lorenz, a meteorologist, in the early 1960s. At that time Lorenz used a sea gull as the disturbing element. He introduced the butterfly in the title of an unpublished paper in 1972, "Does the flap of a butterfly's wings in Brazil set off a tornado in Texas" (Gleick, 1987).

2   The feedback mechanisms are numerous and not easy to evaluate. For example, at a simplistic level an increase in greenhouse gases affecting the radiation balance may increase temperature, which may increase evaporation and thus cloud amount, which in turn should change the radiation balance, etc. The incorporation of all such mechanisms into numerical models and the documentation of their effects through the system is an enormous task.

3   Despite several decades of observation our knowledge of the three dimensional atmosphere is still inadequate especially over the oceans. Our knowledge of the oceans and biosphere, and of their interactions with the atmosphere, is even weaker. The importance of these elements cannot be underestimated. Regardless of the real relationship between sea surface temperature and weather at distant places the El Niño phenomenon has recently caught everyone's attention. The oceans are very different from land. Not only are they always "wet" they have a much greater inertia than the atmosphere so they take much longer to come to equilibrium after some input/output change. Similarly the biosphere plays a very important role. Even its effect on CO_2 exchange with the atmosphere is still open to debate since estimates of the budget of that gas are not balanced.

The development of theory currently used in climatology coincides with the evolution of science in general. Concepts and relationships, that have broad application in fields such as chemistry and physics, have been adapted to the peculiar characteristics of the atmosphere. The early history of atmospheric theory therefore necessarily encompasses the research into gases, mechanics, radiation, and mathematics.

Today it is common to refer to specific relationships or laws by the name of some individual who claimed, or was identified by others, to have originated that expression of the idea. In many situations such naming is legitimate. It is also the case that more than one individual may simultaneously synthesize the information that is available in a given time period. It is seldom, even in early times that research was performed in a vacuum, so several people will be aware of previous knowledge and experiments that point to a particular conclusion. Typically unacknowledged contributions may come from acquaintances, teachers and publications, as well as the individuals who actually developed previous theory and performed critical experiments. Even Newton, in a letter to Hooke in 1675 or 1676, repeated a previously used statement, "If I have seen farther, it is by standing on the shoulders of giants," although one of his biographers, More (1934), questions his motives. Clearly, knowledge at any particular point in time is an accumulation of all previous knowledge. In any one individual it is an incomplete and strongly filtered view. In addition, there are always difficulties in translating past terminology and symbols into current usage. Another problem with attribution is that subsequent work, which may rigorously formulate a given theory, is usually omitted (Boyer, 1939, p. 300). Also, the search to prove some newly identified relationship often produces the most valuable practical applications yet such work is often ignored in historical accounts. Where more than one individual claims authorship to a new idea a long and often acrimonious debate may ensue. The "true" authorship, if indeed only one exists, is frequently impossible to ascertain. In this publication reference will be made only to the generally accepted significant contributors to a concept. Furthermore, it is not the intention here to deal with the nuances of the development of a given theory or to elaborate on its philosophical underpinnings. Rather, the aim is to present a simple chronology so that the reader may gain some sense of the historical background of the science of climatology.

However well a person is prepared in a field at some point derivations appear obscure. It is not just the beginner who has problems in following the logic of others. For example, in the introduction to his translation of Laplace's *Celestial Mechanics* Nathaniel Bowditch (1773–1838) of Salem, MA, stated:

> Whenever I meet in LaPlace with the words "Thus it plainly appears," I am sure that hours, and perhaps days, of hard study will alone enable me to discover *how* it plainly appears. (Laplace, 1825, vol. 1, p. 62)

Unfortunately it is impossible always to go back to the most fundamental of steps in thought (which may be flawed) and this author apologizes to those who have difficulty following his logic.

The philosophy underlying this volume is:

1   The climatology of a region is described by a large number of statistics of the continuum of weather. The period over which the data are sampled is assumed to be sufficiently long to encompass the major features of that climate yet not so long that the controlling factors change. This may be a difficult assumption to fulfill because some factors, such as atmospheric composition, are continuously evolving even without human intervention.

2 Dynamic climatology involves the explanation of climate. By explanation we mean that we may allocate different proportions of the statistics to specific major factors such as latitude, land and sea distributions, orography, etc.

3 The only method presently available for ascertaining the importance of different factors is computer simulation. Currently there are numerous models used for climate prediction. All are gross approximations of reality and they vary enormously in their internal structures. As a consequence, predictions from the same input data differ. Surprisingly, however, the results remain within reasonable bounds and this approach appears to be valid.

4 In order to use such models responsibly a person must be able to understand the underlying principles. These include at a bare minimum, mathematics, statistics, and physics. Other relevant fields are chemistry and biology. In each case several courses are needed to cover the topics in sufficient depth to apply them in climatic modeling.

5 That being stated, we attempt in this volume to outline the background knowledge necessary for a person intending to attempt the explanation of climate. This does not excuse them from taking standard atmospheric courses and reading the thousands of relevant research papers that are the basis of numerical climate prediction models. It is only with a full understanding of such models including their numerous assumptions that we can hope to approach our ultimate goal: understanding climate. This will not be done until significant advances have been made in observation and in the models themselves.

Chapter 2

Mathematics

Mathematics is the language of science and an awareness of its basic principles is fundamental for an understanding of dynamic climatology. Algebra, geometry, and trigonometry may be traced back to the Babylonians and Egyptians but their formal foundations were laid down by the Greek schools following thinkers such as Pythagoras (ca. 580–500 BC), Plato (429–348 BC) and Euclid (ca. 300 BC) (Smith, 1923). Since these subjects are currently assumed to have been covered in pre-college mathematics, that body of knowledge will not be summarized here. It must not be ignored, however, because like the sciences in general, mathematics tends to be a highly sequenced subject and an understanding of more advanced topics requires a good knowledge of the underlying concepts. It is often the lack of a full appreciation of these underlying concepts that gives individuals problems later with mathematics and the sciences that use mathematics extensively. For that reason an outline of some elementary aspects leading to more advanced topics will be given.

Unless otherwise stated, the usual mathematical conventions will be used. In the Cartesian system, x increases to the right or towards the east, y increases up the page or towards the north and z is perpendicular to the x, y plane, out of the page or upwards. This is not always given and many a student has had problems reading Sir Harold Jeffreys' papers that have a different arrangement. Angles are in units of radians (see next section) and are measured counterclockwise from the *abscissa*, the x axis. Therefore, positive rotation around the z axis is the same as the right hand screw rule in vectors: turning a conventional screw, which is pointing in the z direction, clockwise moves the screw towards increasing z. In contrast, in meteorology angles are measured in degrees clockwise from north with the air flow being named from the direction it comes. Thus a west wind has an angle of 270 and a north wind 360 degrees. The angle zero is reserved for calm.

The notation $[j]$ is used to label a sequence, e.g. $x[j]$. Because we often consider the first observation in a sequence as occurring at the origin the first value of j will be 0. Therefore, if there are n observations, the last observation will be $x[n - 1]$ and the jth observation is $x[j - 1]$.

2.1 Geometry

2.1.1 The Circle, Sphere, and Pi

The circle is defined as the locus of a point moving at a given distance, the radius, from a fixed point, the center. If the center is located at a point x_c, y_c and the radius is r then from Pythagoras' theorem

$$(x - x_c)^2 + (y - y_c)^2 = r^2,$$

or

$$x^2 + y^2 - 2x_c x - 2y_c y + x_c^2 + y_c^2 - r^2 = 0. \tag{2.1}$$

For a circle with its center at the origin this becomes

$$x^2 + y^2 = r^2,$$

or

$$\frac{x^2}{r^2} + \frac{y^2}{r^2} = 1 \tag{2.2}$$

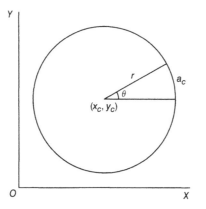

Figure 2.1 The circle with radius r centered at x_c, y_c

Also, for a circle with its center at the origin, from trigonometry

$$x = \cos \theta,$$
$$y = \sin \theta,$$

(2.3)

where the angle θ is measured counter-clockwise from the abscissa.

Estimation of the circumference of the circle, C, from the radius, r, has been a human quest since the earliest times. Both the Babylonians and the Egyptians recognized that the ratio of the circumference to the diameter, $C/2r$, was a constant and had approximated its magnitude by 2,000 BC (Beckmann, 1970). The symbol π was not adopted, however, until the eighteenth century. (For the current magnitudes of the various mathematical and physical constants see appendix B.)

An angle is defined as the ratio of the arc of a circle, a_c, to the radius of that circle, r,

$$\theta = \frac{a_c}{r}.$$

(2.4)

One radian, the unit of angular measure (really no units), is defined by Equation (2.4) when $a_c = r$. Of course, when $a_c = C$, $\theta = 2\pi$, i.e.

$$\pi = \frac{C}{2r}.$$

(2.5)

By analogy a solid angle is defined as the ratio of an area, α, produced by a cone on the the surface of a sphere to the square of the radius, r^2. The vertex of the cone lies at the center of the sphere as in figure 6.2.

$$\omega = \frac{\alpha}{r^2}.$$

(2.6)

One steradian (abbreviated "sr"), the unit of solid angular measure, is defined by Equation (2.6) when $\alpha = r^2$. Again, when $\alpha =$ the surface area of a sphere, $\omega = 4\pi$.

2.1.2 The Ellipse

A most important variable in climate is the earth–sun distance. Most students know that the earth follows an approximately elliptical orbit around the sun and that the degree to which that orbit varies from a circle is given by the *eccentricity*. However few can actually give a good description of this quantity.

The ellipse is defined by a point P moving so that its distance from the focus S is equal to the product of e and the distance from a fixed line, called the *directrix*, where $e < 1.0$ (see figure 2.2), and is called the *eccentricity*.

$$SP = ePM.$$

(2.7)

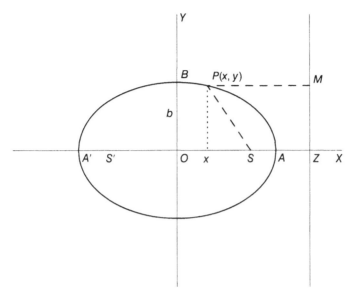

Figure 2.2 The ellipse, $x^2/a^2 + y^2/b^2 = 1$ plotted on axes OX, OY, where $b^2 = a(1 - e^2)$. $e < 1.0$, $e = SP/PM$. S is called the focus, ZM is called the directrix, $OA = a$ is the semi-major axis and b is the semi-minor axis. $OS = ae$ and $OZ = a/e$. As e approaches zero the ellipse approaches a circle and when e approaches 1.0 the ellipse approaches a straight line. In this figure $e = 0.745$

The origin of the axes is obtained by drawing SPM as a straight line, calculating the two possible positions of P along that line, A and A', and bisecting the distance between A and A'.

$$SA = eAZ, \qquad (2.8)$$
$$SA' = eA'Z. \qquad (2.9)$$

Let the semi-major axis $OA = a$, then the addition of Equations (2.8) and (2.9) gives

$$SA + SA' = e(AZ + A'Z),$$
$$2 \times SA = e[(OZ - a) + (OZ + a)],$$
$$2a = e2OZ.$$

Therefore,

$$OZ = \frac{a}{e}, \qquad (2.10)$$

and the equation for the directrix is

$$x = \frac{a}{e}. \tag{2.11}$$

Similarly, Equation (2.9) minus Equation (2.8) leads to

$$OS = ae, \tag{2.12}$$

which gives the location of the focus, $S(ae, 0)$. The other focus is $S'(-ae, 0)$. The equation for an ellipse is obtained from applying Pythagoras' theorem to the triangle SPx,

$$(ae - x)^2 + y^2 = SP^2 = e^2 PM^2 \tag{2.13}$$

$$= e^2 \left(\frac{a}{e} - x \right)^2 \tag{2.14}$$

$$a^2 e^2 - 2aex + x^2 + y^2 = a^2 - 2aex + e^2 x^2 \tag{2.15}$$

$$x^2(1 - e^2) + y^2 = a^2(1 - e^2) \tag{2.16}$$

$$\frac{x^2}{a^2} + \frac{y^2}{a^2(1 - e^2)} = 1. \tag{2.17}$$

Let $b^2 = a^2(1 - e^2)$, then

$$\frac{x^2}{a^2} + \frac{y^2}{b^2} = 1, \tag{2.18}$$

is the equation of an ellipse. It was noted in the caption to figure 2.2 that as e approaches zero the ellipse approaches a circle (Equation 2.2) and when e approaches 1.0 the ellipse approaches a straight line.

2.2 Differential Calculus

Since the field of dynamics deals with constantly varying quantities we need a set of tools that will handle change. The major branch of mathematics originally developed for just this purpose is calculus. It deals with the instantaneous variation of elements such as rates of change of distance with respect to time which we call speed, as in miles or kilometers per hour. Notice that the word "per" may be translated as "divided by." Thus 50 kilometers per hour $=$ 50 km/1 hour $=$ 50 km hr^{-1}. That means, assuming we travel at a constant speed, that we shall go 50 km in one hour. This is illustrated in figure 2.3 on a distance verses time plot. We start out at P, distance y_1 miles along the road at time t_1 hours and arrive at Q, y_2 along the road at time t_2 hours.

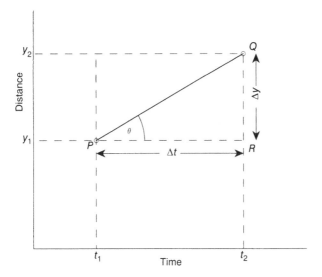

Figure 2.3 Speed as a function of time and distance

The speed is given by

$$\text{Speed} = \frac{(y_2 - y_1)}{(t_2 - t_1)} = \frac{\Delta y}{\Delta t}. \tag{2.19}$$

The Greek letter delta is sometimes written in the upper case, Δ, and sometimes, as below, by the lower case, δ. They are used to represent the same thing, "a finite difference in." The ratio of the distances, $QR/PR = \Delta y/\Delta t = \delta y/\delta t$, also represents the trigonometric function known as the tangent of angle θ.

Of course, we seldom travel at a constant speed and, although we typically use an hour for the unit of time, we are really interested in the speed at an instant (i.e. the speedometer usually displays a continuously varying speed). An instant presumably involves zero time. But this presents a dilemma because the distance traveled in zero time must be zero and the fraction 0/0, giving speed, is undefined. Only if we allow time to approach zero without actually becoming zero will the fraction have meaning and the speed will approach a limiting magnitude, what we call the *instantaneous* speed. The mathematical process is called *differential calculus* or *differentiation*.

As is typical of many discoveries, the original discussion of a topic often seems tortuous and difficult. Subsequent re-formulations in the light of further

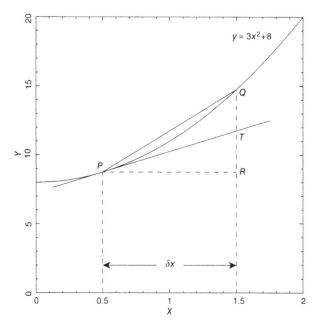

Figure 2.4 Chord PQ and tangent PT on $y = 3x^2 + 8$

knowledge and development of notation do, however, make explanations appear simpler. Such is the case with differentiation which is now relatively easily explained. The major step in the development of calculus was the introduction of the concept of a limit and of techniques for obtaining solutions for that limit.

In this case we are attempting to obtain the gradient of a curve at any point, where the gradient is defined as the slope of the tangent line to the curve at that point. The curve in a simple two-dimensional case is defined by an equation involving two variables. As a specific example this might be $y = 3x^2 + 8$. In general this equation may be written as $y =$ function of x, or $x = f(x)$. The latter form of notation was first introduced by Euler in 1734 although he did not published it until 1740 (Youschkevitch, 1981).

The process of calculating the tangent PT involves the drawing of a chord through the point P, the location where the gradient is required, to a neighboring point Q (figure 2.4). The slope of this chord is the average gradient between P and Q and may be stated

$$\textit{Slope of PQ} = \frac{QR}{PR} = \frac{\delta y}{\delta x}$$

which is equivalent to Equation (2.19). In general terms this may be written as

$$Slope\ of\ PQ = \frac{f(x + \delta x) - f(x)}{\delta x},$$

where δy is replaced by the function of x.

For the specific function, $y = 3x^2 + 8$, where y might represent distance and x represents time, the slope is

$$\frac{[3(x + \delta x)^2 + 8] - [3x^2 + 8]}{\delta x} = \frac{[3x^2 + (3)(2)x\delta x + 3\delta x^2 + 8] - [3x^2 + 8]}{\delta x}$$

$$= \frac{(6x + 3\delta x)\,\delta x}{\delta x}. \tag{2.20}$$

And at the particular point P at $x = 0.5$, this becomes

$$\frac{QR}{PR} = \frac{(3 + 3\delta x)\,\delta x}{\delta x}. \tag{2.21}$$

Now Q is allowed to approach P, i.e. δx is made to approach zero without actually being set to zero. The limiting magnitude of the slope of the chord as $Q \to P$ is the slope of the tangent, which may be written as

$$\lim_{\delta x \to 0} \frac{\delta y}{\delta x}.$$

In the example above, so long as δx is not zero the δx's may be canceled leaving $(3 + 3\delta x)$. Furthermore, if we make δx as small as we please this becomes 3, which is the slope of the tangent of the function $y = 3x^2 + 8$ at $x = 0.5$. When y and x represent distance and time respectively the slope of 3 would have the units of speed. This is called the first derivative of the function $y = f(x)$. The first derivative is sometimes represented by $D_x y$, sometimes by $f'(x)$, but more usually by dx/dy. The symbol "D" was first used as a differential operator by Louis Arbogast (1759–1803) but it was popularized by Heaviside in the late nineteenth century (see section 2.6.4).

Thus, in current notation, the first derivative may be written

$$\frac{dy}{dx} = \lim_{\delta x \to 0} \frac{f(x + \delta x) - f(x)}{\delta x} = f'(x). \tag{2.22}$$

It should be noted that dx, the differential of x, might represent any magnitude change in x. Then the element dy, however, is defined by as the product of $f'(x)$ and dx. Thus in figure 2.4, if $dx = \delta x$, then dy, as part of the derivative, is RT not RQ. This sometimes causes confusion especially when dy/dx is used as if it were a fraction.

General formulae for derivatives of powers of x, of products, and of quotients are shown to be

$$\frac{dx^n}{dx} = nx^{n-1},$$ (2.23)

$$\frac{d(uv)}{dx} = u\frac{dv}{dx} + v\frac{du}{dx},$$ (2.24)

and

$$\frac{d(u/v)}{dx} = \frac{v\dfrac{du}{dx} - u\dfrac{dv}{dx}}{v^2}$$ (2.25)

respectively. The variables, u and v, are each functions of x. A constant has no slope so the derivative of a constant is zero. Other derivatives may be found in calculus textbooks and tables of mathematical functions, such as Abramowitz and Stegun (1964), Korn and Korn (1967), or Selby (1967).

Where y is a function of z, which is a function of x, the *chain rule* may be applied to find the derivative of y with respect to x:

$$\frac{dy}{dx} = \frac{dy}{dz} \times \frac{dz}{dx}.$$ (2.26)

Differentiation may be repeated on the derivative function to produce the second derivative, and in turn to produce the third, etc. To represent the nth derivative the following is used

$$\frac{d^n()}{dx^n} = f^n().$$ (2.27)

2.3 Partial Derivatives

Frequently a variable, say z, is a function of more than one independent variable, $z = f(x, y)$. For example, the temperature of the sea surface is a function of x and y, the distances to the east and north of the origin in the Cartesian system. So long as the independent variables are not functions of one another, i.e. they are said to be orthogonal, differentiation may proceed with respect to each variable independently. This is known as partial differentiation and is represented by the curly d, "∂".

In figure 2.5 the variable z is given by

$$z = \left(1 - \frac{x^2}{4}\right)^{\frac{1}{2}} + (1 - y^2)^{\frac{1}{2}}$$ (2.28)

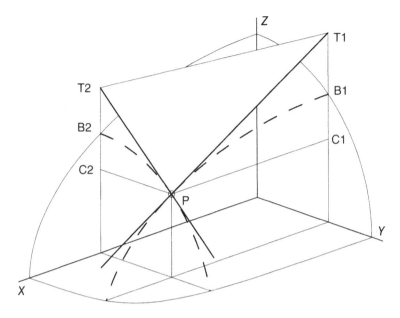

Figure 2.5 Tangent plane on curved surface

which is the sum an ellipse in the zx plane and a circle in the zy plane. Each term in this equation contains only an x or a y but there is no constraint on there being complex combinations of x and y. At the point P, two tangents in the perpendicular planes may be obtained. If y is kept constant the tangent in the zx plane is drawn as P–T1 and if x is kept constant the tangent in the zy plane is P–T2. The slopes in the different coordinate directions arecalculated in the usual way to give

$$\left(\frac{dz}{dx}\right)_{y\ constant} = \frac{\partial z}{\partial x} = -\frac{1}{4}\frac{x}{\left(1 - \frac{x^2}{4}\right)^{\frac{1}{2}}}, \quad \text{and} \tag{2.29}$$

$$\left(\frac{dz}{dy}\right)_{x\ constant} = \frac{\partial z}{\partial y} = -\frac{y}{(1 - y^2)^{\frac{1}{2}}}, \quad \text{respectively.} \tag{2.30}$$

Figure 2.6 illustrates the relationship between the total differential of z, dz, and the partials. From the triangle P–C1–T1

$$\left(\frac{\partial z}{\partial x}\right)_{at\ P} = \frac{z_2 - z_1}{x - x_1}, \quad \text{i.e. } (z_2 - z_1) = \left(\frac{\partial z}{\partial x}\right)_{at\ P}(x - x_1). \tag{2.31}$$

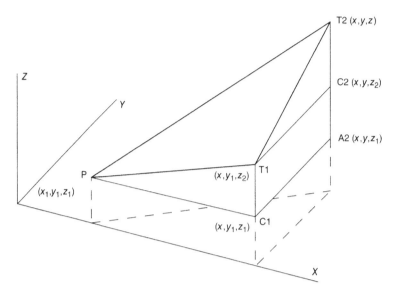

Figure 2.6 Coordinates for two tangents P–T1 and P–T2

Similarly from triangle T1–C2–T2

$$(z - z_2) = \left(\frac{\partial z}{\partial y}\right)_{at\ P} (y - y_1). \tag{2.32}$$

Add these two equations and

$$(z - z_1) = \frac{\partial z}{\partial x}(x - x_1) + \frac{\partial z}{\partial y}(y - y_1) \tag{2.33}$$

which is the equation of the tangent plane at P. The quantity $(z - z_1)$ is the total differential, dz, on that plane. In general then, the total differential is

$$dz = \frac{\partial z}{\partial x} dx + \frac{\partial z}{\partial y} dy, \tag{2.34}$$

where the partials are obtained from the particular function [in the case of figure 2.5 given by Equations (2.29) and (2.30)]. Division of this equation by dx reveals that the total derivative of z with respect to x is equal to the partial derivative of z with respect to x plus the product of the partial derivative of z with respect to y and the total derivative of y with respect to x.

Equation (2.34) may be further generalized for several variables. Thus if $s = f(t, x, y, z)$

$$ds = \frac{\partial s}{\partial t} dt + \frac{\partial s}{\partial x} dx + \frac{\partial s}{\partial y} dy + \frac{\partial s}{\partial z} dz. \qquad (2.35)$$

Division this time by dt gives

$$\frac{ds}{dt} = \frac{\partial s}{\partial t} + \frac{\partial s}{\partial x}\frac{dx}{dt} + \frac{\partial s}{\partial y}\frac{dy}{dt} + \frac{\partial s}{\partial z}\frac{dz}{dt}, \qquad (2.36)$$

which, as we shall see later, is used frequently in atmospheric calculations.

2.4 Integral Calculus

The previous pages have given a brief review of differential calculus. Quite a different example dealing with rates of change might involve the calculation of the distance traveled given the speed. Let us assume that we travel 5 km hr^{-1} for 12 min, 10 km hr^{-1} for 18 min and 20 km hr^{-1} for 30 min. The total distance traveled is obtained from multiplying each speed with the respective time at that speed and then adding, or summing, the results together:

$$\frac{5\,km}{60\,min} \times 12\,min + \frac{10\,km}{60\,min} \times 18\,min + \frac{20\,km}{60\,min} \times 30\,min = 14\,km.$$

Thus the total distance traveled is 14 km. Graphically this is shown in figure 2.7. With axes, speed for the ordinate (vertical on the page) and time for the abscissa (horizontal on the page), distance is the area within the graph. In a real situation the speed would not have remained constant for long periods: it would have continuously varied with time. Nevertheless, the process of calculating distance would have been the same. Each speed would be multiplied by a very small interval of time (approaching the limit of zero) and the results added together. The mathematical process is called *integral calculus* or *integration*. It is the reverse of the differential process.

It should be noted that, while the variables in the example in figure 2.7 were distance and time with their product giving speed, there is no restriction on the process to these particular variables: any variables that are related may be operated upon in these ways.

Again, current knowledge permits easy explanation (here without proof) of integration. As indicated above, distance is obtained from speed by the multiplication of speed by time, or from the area on a speed verses time plot. In a specific example shown in figure 2.8 let speed, y, be given by the relationship $y = 2x^2 + 12$, where x is time. The total distance traveled is proportional to the area between the curve representing speed, the time axis, X, and the beginning and ending times "a" and "b". At P for an interval of

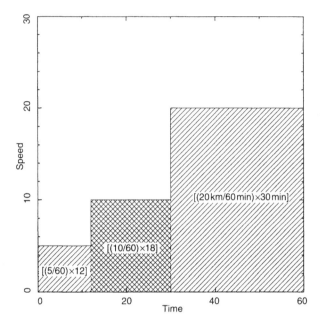

Figure 2.7 Areas representing distances traveled

time δx, the distance traveled, δS, is given by an area in magnitude between $y \, \delta x$ and $(y + \delta y) \, \delta x$, i.e. the rectangles pPRq and pUQq. This may be written

$$y\delta x < \delta S < (y + \delta y) \, \delta x$$
$$y < \frac{\delta S}{\delta x} < (y + \delta y).$$

But as $\delta x \to 0$, $\delta y \to 0$, and

$$\lim_{\delta x \to 0} \frac{\delta S}{\delta x} = \frac{dS}{dx} = y.$$

Or,

$$\delta S = \lim_{\delta x \to 0} y \, \delta x.$$

Clearly, the area S is the sum of all the small areas $y \, \delta x$, with these small areas being accurate representations as $\delta x \to 0$,

$$S = \lim_{\delta x \to 0} \sum y \, \delta x = \int y \, \delta x \tag{2.37}$$

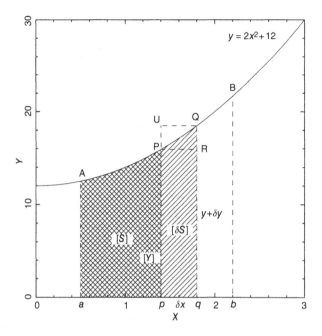

Figure 2.8 Areas representing distances relating to the equation for speed $y = 2x^2 + 12$ where x is time

where the symbol "$\int$" is a long "s" known as the *integral* sign. The right hand side of Equation (2.37) may be read as "integrate y with respect to x." The question now is, "What is S and how may it be obtained?" From Equation (2.23), if we replace n with $n + 1$, we see that $d(x^n)/dx = nx^{n-1}$. Therefore

$$\frac{d\left(\dfrac{1}{n+1}x^{(n+1)}\right)}{dx} = x^n. \tag{2.38}$$

Thus, if the example in figure 2.37 is used ($y = 2x^2 + 12$), one answer is

$$\frac{d\left(2\dfrac{1}{(2+1)}x^{(2+1)} + 12\dfrac{1}{(0+1)}x^1\right)}{dx} = 2x^2 + 12.$$

In this case

$$S_{first\ estimate} = \frac{2}{3}x^3 + 12x. \tag{2.39}$$

Also, as we saw with differentiation, a constant, which moves the curve vertically but does not affect the gradient, disappears. Therefore the full answer is

$$S_{general\ solution} = \frac{2}{3}x^3 + 12x + c. \tag{2.40}$$

In practice we are usually interested in the distance traveled during a particular time period "a" to "b". In which case we perform what we call *definite integration*. This involves evaluating the function S at "b" and subtracting the value at "a". The constant "c" then disappears.

$$S = \int_a^b y\,\delta x = \left[\frac{2}{3}x^3 + 12x\right]_a^b = \left[\frac{2}{3}(b^3 - a^3) + 12(b - a)\right]. \tag{2.41}$$

For the values of $a = 0.5$ and $b = 2.2$ displayed in figure 2.8, $S = 27.1453$. What is really being calculated at a specific x, i.e. at b, is the whole area from $x = 0$ to $x = b$. The area between a and b is then the difference of the two totals. For this to be possible the function y must be defined uniquely over the whole interval: the path of y must be known. In some cases in thermodynamics (see section 5.5.2) we shall find that this is not necessarily the case.

The general equation for the integral of simple powers of x is

$$\int (kx + m)^n\,dx = \frac{(kx + m)^{(n+1)}}{a(n + 1)} + c \tag{2.42}$$

with the exception of x^{-1} whose integral is the Naperian logarithm x.

Again, more complex integrals may be found in calculus textbooks and tables of mathematical functions as referenced at the end of section 2.2.

2.5 Development of Calculus

As has been seen, central to the development of calculus is the idea of a limit both for differentiation and for integration. This was not resolved until the seventeenth century. By that time a large body of knowledge existed on geometrical problems and their expressions in series. Much was contributed by the French school. For example, in 1593 François Viète (1540–1603) had introduced the "..." notation for infinite series. Also, René du Perron Descartes (1596–1650), while he lived in Holland, established the link between geometry and algebra. It was in the appendix to his book on *Discourse on Method* published in 1637 that he outlined analytical geometry through which he represented curves by algebraic equations (Roth, 1937). He used the idea of latitude and longitude to uniquely locate a point in a plane with the coordinates x and y. He applied the term *ordinate* from the Latin lineæ ordinatæ meaning parallel lines. [The word *abscissa* was first used by Stefano degli Angeli in Italy in 1659 – (Cajori, 1893).] Although the rectangular coordinate system of representation is now known as the *Cartesian system* after Descartes, he used various oblique axes that he found appropriate to the problem. Fellow Frenchmen, Giles Persone de Roberval (1602–75) and Pierre de Fermat (1601–65) each developed methods for drawing tangents with the latter giving the solution for finding maxima and minima of curves. Roberval also worked on the method of indivisibles, i.e. an infinite number of infinitely narrow rectangles. This was applied to the problem of obtaining the area under a curve (integration), then known as *quadrature* or *squaring*. It was closely related to the technique of *exhaustion* of Archimedes, (i.e. the use of exterior and interior rectangles, pPRq and pUQq, with pq made smaller and smaller as shown in figure 2.8). However, without the concept of a limit such an approach did not always give a solution, as it did not with $y = 1/x$, the hyperbola.

Across the Channel more mathematicians were at work. In 1631 in England William Oughtred (1574–1660) published *Calvis mathematicæ* that contained essentially all that was known about arithmetic and algebra at the time. He invented the slide ruler, both in linear and circular form, and was responsible for making the research of John Napier more widely known (see section 2.7.1). He introduced "::" for "proportional to", and "∼" for "difference between." He also used × for multiplication which Leibniz thought to be a bad choice because it was too much like "*x*." John Wallis (1616–1703), professor of geometry at Oxford from 1649, studied infinitesimals and quadratures. He was the first to use the symbol "∞" for "infinity." A contemporary, Isaac Barrow (1630–77), was appointed at Cambridge first as professor of Greek in 1660 and then as professor of mathematics in 1663, a post he gave up for Newton in 1670. He also worked on infinitesimals and the calculation of the tangent.

It was into this mathematical environment that Isaac Newton (1642–1727) was thrust in 1660 when he entered Trinity College, Cambridge. Besides being taught by Barrow he studied Euclid, Descartes, Oughtred and Wallis amongst others. He received his Bachelor of Arts in 1665 and an MA in 1668. At two times, August 1, 1665 to March 25, 1666 and June 22, 1666 to March 27, 1667, the university was closed because of the plague and Newton had to return to his Woolsthorpe, Lincolnshire home. It was apparently during this period that his original ideas on calculus, mechanics, and optics evolved. Unfortunately, he did not immediately put them into print. His early papers on optics were published in the *Philosophical Transactions of the Royal Society* between 1672 and 1676 and his work on mechanics in *Principia*, in 1687. However, with the exception of discussions in *Principia* (Newton, 1687) and in an appendix to *Opticks* (Newton, 1704) no real independent publication of Newton's calculus appeared until 1737 (Whiteside, 1964). As a result, a long and bitter feud erupted between Newton and Leibniz and their followers with regard to which one invented the calculus. Notes, manuscripts and material contained in letters now reveal that Newton's claim to the 1665–6 date is legitimate.

Besides not immediately publishing his research, Newton further impeded the adoption of his ideas by changing his notation and selecting notation that was not easily set up for printing. Newton was aware of this problem, at least later in his life, for in writing of himself in an unsigned article in 1717 (Cajori, 1929, p. 200) he stated, "Mr Newton doth not place his method in Forms of Symbols, nor confine himself to any particular sort of Symbols for Fluents and Fluxions." The term *fluent* referred to the variable that was considered to be a function of time and *fluxion* to the time derivative of that variable. Cajori points out that Newton used "pricked" letters to represent fluxions as early as May 20, 1665 (Cajori, 1929, p. 197). Thus

$$\dot{x}, \dot{y}, \ddot{x} \text{ represented } \frac{dx}{dt}, \frac{dy}{dt}, \frac{d^2x}{dt^2} \text{ respectively.}$$

For example, the fluxion of the equation in figure 2.8, $y = 3x^2 + 8$ becomes

$$\dot{y} = 3 \times 2x\dot{x} \tag{2.43}$$

Such notation was convenient for time derivatives and was adopted by many throughout the following centuries. In fact, it has also been used in Equations (4.16) and (4.17) below. The form for the differentiation of radicals and fractions became much more complex and cumbersome. Also, to denote derivatives with respect to variables other than time, two fluxions and a ratio sign were necessary:

$$\dot{z} : \dot{x} = \frac{dz}{dt} : \frac{dx}{dt} = \frac{dz}{dx}.$$

The use of the dot did not originate with Newton since it had been used previously by Digges in 1592 to represent small quantities and by Mercator in 1668 (before Newton's notation was introduced by Wallis in 1693) for infinitesimals. Making his notation even more difficult to follow, Newton often used completely different letters, such as p, q and r, and even line segments such as DE, FG and HI, for the fluxions of x, y and z. In *Principia* in Book II Lemma II Newton did use matching lower case letters to represent what he called moments of a variable: "the moment of any power $A^{\frac{n}{m}}$ will be $\frac{n}{m} a A^{\frac{n-m}{m}}$."

For "infinitely little Quantities" he sometimes used "o." Thus in his proof of differentiation in *A Treatise of the Method of Fluxions and Infinite Series*, originally written in 1671 but not published until 1737, he essentially used Equation (2.22) where δx was called the moment of x and was represented by $\dot{x}o$. He was then able to use the binomial theorem, which he had solved for negative integers and fractions, to expand $(x + \dot{x}o)^n$ and give the general solution for the derivative. As has already been indicated above, the derivative of y with respect to x was obtained by dividing the time derivative of y by the time derivative of x.

Newton also recognized that derivatives and integrals were the inverses of one another. As an example for the integral Newton (1704) used the rectangle, $\boxed{\frac{aa}{64x}}$ for the current $\int \frac{aa}{64x} \, dx$.

Despite his reticence to publish, Newton's genius was widely acknowledged early in his career. He was appointed Lucasian professor at Cambridge in 1667 at age 26. A requirement of this position was to give copies of his lectures to the library. As a result, manuscripts on optics (1670–2), arithmetic and algebra (1673–83), and much of *Principia* (1684–7) were deposited. He became a member of Parliament in 1689 and again in 1701–2 and was appointed Warden of the Mint in 1696 and Master in 1699 but he did not resign his professorship until 1701. He was President of the Royal Society from 1703 until his death in 1727. Queen Anne gave him a knighthood in 1705.

Gottfried Wilhelm Leibniz (1646–1716), born in Leipzig, went to the university there at age 15 to study law. He completed his degree at Jena and Altdorf, graduating in 1666. In service to the archbishop elector of Mainz from 1667–76 he spent the years of 1672–6 traveling. In Paris he met Huygens and studied the works of the French mathematicians including Descartes and Pascal. In London he met John Collins who had received a copy of Newton's unpublished manuscript *De Analysi per Æquationes* from Isaac Barrow in 1669. It is unknown whether Leibniz saw this document. Newton and Leibniz subsequently in 1676, corresponded. In 1684 Leibniz published his first paper on differential calculus in *Acta Eruditorum* (also known as the Leipzig Acts). He provided no proof but he presented the derivatives for products and quotients in exactly the same notation and form as in Equations (2.23), (2.24), and (2.25) above. According to Cajori (1929) Leibniz had used dy/dx in a manuscript as early as 1675 when he also used the long "s," $\int$, to represent "sum of" in place of the letters "omn," the Latin "all." The symbol "$\sum$" for "sum of" was not introduced until 1755 by Leonhard Euler. The *Acta Eruditorum* paper was discovered by the Swiss, Jacob Bernoulli in 1687. Jacob (1654–1705) was from a large family of Bernoullis many of whom made significant contributions to mathematics and science. He and his brother Johann (1667–1748) did much to develop calculus more fully. In 1690 Jacob used the term *integral* for the first time and in 1696 Leibniz and Johann agreed to replace the term "calculus summatorius" with "calculus integralis." The Bernoullis were staunch supporters of Leibniz in his claim to discovering the calculus before Newton. For much of the latter part of his life Leibniz worked on his philosophy concerning theology.

Whereas it is now clear that Newton's discovery of the calculus predates that of Leibniz, it is Leibniz's notation and rules (Equations 2.23 to 2.26) that were generally adopted and played an important role in the subsequent development of that area of mathematics, especially on the continent of Europe in the following century.

Although Newton had employed partial differentials in the 1660s, and contemporary mathematicians such as Leibniz and later the Bernoullis used them, no special notation was adopted. Legendre introduced the curly d in its present form in 1786 but it was ignored until C. G. J. Jacobi applied "d" for the "total" and "∂" for the partial derivatives in 1841 (Cajori, 1929).

2.6 Vectors

So far we have considered variables that are termed *scalars* like speed, time, and distance. These quantities only have magnitudes and are represented by letters or symbols in normal font. Other variables with which we shall be much concerned, such as those dealing with flow, also have the property of direction. Their description therefore requires a pair of magnitudes. These are called *vectors*. Wind, the movement of air, has a speed and a direction. Such pairs are often represented by letters or symbols in bold font or with a bar or arrow over them, e.g. $\mathbf{V}$, $\overline{V}$, or $\vec{V}$. Bolding, introduced in the nineteenth century, is convenient for printing but the overbar, originally used by Argand in 1806, is more easily added in handwriting. Figure 2.9 displays a three-dimensional vector in Cartesian coordinates.

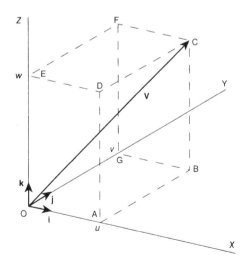

Figure 2.9 Vector $\mathbf{V}$, components u, v, and w, and unit vectors $\mathbf{i}$, $\mathbf{j}$, and $\mathbf{k}$

2.6.1 Addition and Subtraction

Addition may be applied in the usual way but multiplication is different. Graphically, addition is performed by placing the tail of a second vector on the head of the first. The sum is given by the head of the second and the new vector drawn from the tail of the first to the head of the second. Subtraction is performed in a similar way with the negative vectors reversed, head to tail. Rather than operating on the original vectors there are often advantages in subdividing them into orthogonal components. In the Cartesian system this is illustrated in figure 2.9. Perpendiculars from the head of the vector **V** to the three coordinate planes, xy, yz, and zx, create a box, ABCDEFGO with **V** as the diagonal OC. The components are then given as "u" for the x axis, "v" for the y axis and "w" for the z axis. Sometimes these are labeled V_x, V_y, and V_z, respectively. Trigonometry provides a method for calculating these components. For example, from the magnitude of V and the angle BOA in the xy plane $u = V \sin(\text{BOA})$, $v = V \cos(\text{BOA})$.

Once all vectors have been resolved into their components each coordinate may be operated upon separately as if each element were a scalar. This often simplifies calculations. If required, the components may be combined into their "resultant" vectors after processing. Since the components are still vectors it is often necessary to identify them as such. For this purpose the unit vectors, **i**, **j**, and **k**, in the x, y, and z directions respectively, are introduced (see figure 2.9). Then the wind component in the x direction is **i**u. Thus we may represent **V**$_1$ as **i**u_1 + **j**v_1 + **k**w_1 and **V**$_2$ as **i**u_2 + **j**v_2 + **k**w_2 and addition and subtraction becomes

$$\mathbf{V}_1 \pm \mathbf{V}_2 = \mathbf{i}(u_1 \pm u_2) + \mathbf{j}(v_1 \pm v_2) + \mathbf{k}(w_1 \pm w_2) \qquad (2.44)$$

where either all positive or all negative signs are selected from "$\pm$."

2.6.2 Multiplication

There are two types of vector multiplication. One, called *dot multiplication*, results in a scalar. It is defined as the product of the magnitudes of the vectors and of the cosine of the angle between them,

$$\mathbf{V}_1 \cdot \mathbf{V}_2 = V_1 V_2 \cos(\theta). \qquad (2.45)$$

The cosine factor transforms one vector into its component that is parallel to the other. It is used, for example, in the calculation of work, defined as the product of force and the distance moved (see section 4.3.5). Since $\mathbf{i} \cdot \mathbf{i} = \mathbf{j} \cdot \mathbf{j} = \mathbf{k} \cdot \mathbf{k} = 1 \times 1 \times \cos(\pi/2) = 1$, and all other dot products of involving **i**, **j**, or **k** are zero,

$$\mathbf{V}_1 \cdot \mathbf{V}_2 = u_1 u_2 + v_1 v_2 + w_1 w_2. \qquad (2.46)$$

In the second type of multiplication, called *cross multiplication*, the result of the product of two vectors is a third vector pointing in the direction dictated by the "right hand" rule. Its magnitude is given by the product of the magnitudes of the vectors and the sine of the angle between them measured clockwise from V_1 to V_2

$$\|V_1 \times V_2\| = V_1 V_2 \sin(\theta), \tag{2.47}$$

where the vertical lines indicate "magnitude of." The sine factor transforms one vector into its component which is perpendicular to the other. The right hand rule specifies that the resultant vector points into the plane in which V_1 and V_2 lie when viewed so that the vector listed after the cross, V_2, is clockwise from the one listed before, V_1. Because sine is negative between π and 2π, when the angle from the first to the second is greater than π, the resultant vector points out of that plane.

Since $i \times i = j \times j = k \times k = 0$, $i \times j = k$, $j \times k = i$, $k \times i = j$, $j \times i = -k$, $k \times j = -i$, and $i \times k = -j$, (i.e. cross multiplication is not commutative),

$$V_1 \times V_2 = i(v_1 w_2 - v_2 w_1) + j(u_2 w_1 - u_1 w_2) + k(u_1 v_2 - u_2 v_1). \tag{2.48}$$

The cross product may also be expressed as a determinant,

$$V_1 \times V_2 = \begin{vmatrix} i & j & k \\ u_1 & v_1 & w_1 \\ u_2 & v_2 & w_2 \end{vmatrix}. \tag{2.49}$$

The cross product is typically used where rotation is involved. The cross product of an angular velocity (e.g. of a mass circulating at the end of the string) with a position vector (length of string) produces a linear velocity (of the mass). The angular velocity is represented by a vector pointing in a direction according to the right hand rule. Therefore in figure 2.10 the angular velocity of the earth points in the direction from the south to the north pole with a magnitude of $2 \times \pi/$(seconds in a sidereal day) $= \Omega = 7.292 \times 10^{-5}$ radians/second. Then the linear velocity is given by

$$V = \Omega \times r. \tag{2.50}$$

As an example, for point P, at latitude $\phi = 40°$, $V_e = \|\Omega \times r\| = \Omega r \sin(90 - 40°) \approx 356 \,\mathrm{m\,s^{-1}}$ from west to east where $r = 6.371 \times 10^6$ m is the radius of the earth. The position vector lies along a line from the center of the earth towards P. The rotation vector, displayed using the convention of pointing in the direction of a right hand screw, is placed at the north pole but it could be located anywhere. The tangential velocity vector is shown at P where it applies. The quantity, $R = r \sin(90 - 40°)$, is the distance from P to the

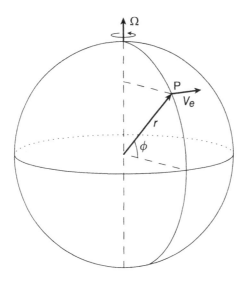

Figure 2.10 Linear velocity at P from $\Omega \times \mathbf{r}$

earth's axis, the radius of rotation for latitude ϕ. The magnitude of $\mathbf{V}_e$, V_e, is ΩR. Thus we have three expressions of the connection between angular velocity, radius of curvature and tangential, or linear, velocity.

$$V = \Omega R, \quad \Omega = \frac{V}{R}, \quad \text{and} \quad R = \frac{V}{\Omega}. \tag{2.51}$$

In meteorology several combinations of multiple products are used. As an example, we meet $\mathbf{A} \times (\mathbf{A} \times \mathbf{B})$. From figure 2.11 we see that the result is a vector perpendicular to $\mathbf{A}$ in the same plane as $\mathbf{A}$ and $\mathbf{B}$ but pointing away from $\mathbf{B}$. The first cross multiplication, $(\mathbf{A} \times \mathbf{B})$, points perpendicular to $\mathbf{A}$ and $\mathbf{B}$ with a magnitude of $AB \sin \theta$. The final vector, $\mathbf{A} \times (\mathbf{A} \times \mathbf{B})$, as a result of the second cross multiplication points perpendicular to $\mathbf{A}$ and $(\mathbf{A} \times \mathbf{B})$ with a magnitude of $A^2 B \sin \theta$.

Newton made considerable use of vectors but the algebra of their use was not developed until the nineteenth century. A significant contributor to this field was William Rowan Hamilton (1805–65). Born in Dublin, he was raised and educated by an uncle. By the age of nine he was supposedly proficient in a dozen languages and was also interested in mathematics and science. He read Newton's *Principia* and found an error in Laplace's work. He attended Trinity College, Dublin, and was appointed astronomer royal and professor there in 1827. He was knighted in 1835. He was much interested in the work of Immanuel Kant but his major contributions were in the fields of algebra, optics, and dynamics. Of particular relevance here is the theory he developed

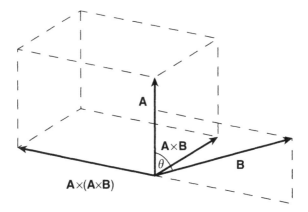

Figure 2.11 Illustration of two cross products

in 1843 on quaternions. This subject, involving four scalars and three vectors, is no longer an important area of study, but within it Hamilton did much to clarify information on vectors.

2.6.3 Differentiation

Differentiation of a vector proceeds in the same way as usual for a fixed coordinate system,

$$\frac{d\mathbf{A}}{dt} = \mathbf{i}\frac{dA_x}{dt} + \mathbf{j}\frac{dA_y}{dt} + \mathbf{k}\frac{dA_z}{dt}. \tag{2.52}$$

For a vector, $\mathbf{A}$, in a rotating system, the unit vectors are variable (in direction not magnitude) and the differentiation of the product then gives

$$\frac{d\mathbf{A}}{dt} = \mathbf{i}_r\frac{dA_{x,r}}{dt} + A_{x,r}\frac{d\mathbf{i}_r}{dt} + \mathbf{j}_r\frac{dA_{y,r}}{dt} + A_{y,r}\frac{d\mathbf{j}_r}{dt} + \mathbf{k}_r\frac{dA_{z,r}}{dt} + A_{z,r}\frac{d\mathbf{k}_r}{dt},$$

$$= \left(\mathbf{i}_r\frac{dA_{x,r}}{dt} + \mathbf{j}_r\frac{dA_{y,r}}{dt} + \mathbf{k}_r\frac{dA_{z,r}}{dt}\right)$$

$$+ \left(A_{x,r}\frac{d\mathbf{i}_r}{dt} + A_{y,r}\frac{d\mathbf{j}_r}{dt} + A_{z,r}\frac{d\mathbf{k}_r}{dt}\right).$$

$$\left(\frac{d\mathbf{A}}{dt}\right)_{fixed} = \left(\frac{d\mathbf{A}}{dt}\right)_{relative} + rotation\ factor. \tag{2.53}$$

Now, since a unit vector may change only with rotation, it maintains a constant magnitude of 1. In figure 2.12 the unit vectors $\mathbf{i}_r$ and $\mathbf{j}_r$ are shown rotating

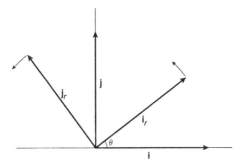

Figure 2.12 Unit vectors $\mathbf{i}_r$ and $\mathbf{j}_r$ rotating with respect to $\mathbf{i}$ and $\mathbf{j}$

through angle θ. Each may be represented by the fixed unit vectors $\mathbf{i}$ and $\mathbf{j}$, i.e. projection of $\mathbf{i}$ and $\mathbf{j}$ onto the $\mathbf{i}_r$ and $\mathbf{j}_r$ axes,

$$\mathbf{i}_r = \mathbf{i}\cos\theta + \mathbf{j}\sin\theta, \tag{2.54}$$

and

$$\mathbf{j}_r = -\mathbf{i}\sin\theta + \mathbf{j}\cos\theta. \tag{2.55}$$

Then differentiation of equations (2.54) and (2.55) gives

$$\frac{d\mathbf{i}_r}{dt} = -\mathbf{i}\sin\theta\,\frac{d\theta}{dt} + \mathbf{j}\cos\theta\,\frac{d\theta}{dt} = \mathbf{j}_r\,\frac{d\theta}{dt}, \tag{2.56}$$

and

$$\frac{d\mathbf{j}_r}{dt} = -\mathbf{i}\cos\theta\,\frac{d\theta}{dt} - \mathbf{j}\sin\theta\,\frac{d\theta}{dt} = \mathbf{i}_r\,\frac{d\theta}{dt}. \tag{2.57}$$

These show that differentiation of a rotating unit vector is a vector perpendicular to itself with a length equal to the rate of rotation. From (2.57), since $d\theta/dt = \Omega$,

$$\frac{d\mathbf{i}_r}{dt} = \Omega \times \mathbf{i}_r. \tag{2.58}$$

Similarly,

$$\frac{d\mathbf{j}_r}{dt} = \Omega \times \mathbf{j}_r, \tag{2.59}$$

and

$$\frac{d\mathbf{k}_r}{dt} = \Omega \times \mathbf{k}_r. \tag{2.60}$$

The relationship between the differentiation of a vector in a fixed coordinate system and a relative (rotating) system (Equation 2.53) may now be rewritten with substitution from (2.58), (2.59), and (2.60):

$$\left(\frac{d\mathbf{A}}{dt}\right)_{fixed} = \left(\frac{d\mathbf{A}}{dt}\right)_{relative} + \mathbf{\Omega} \times \mathbf{A}. \tag{2.61}$$

2.6.4 Differential Operator

William Hamilton also introduced the vector differential operator with the symbol ▷, which later became ∇,

$$\nabla = \mathbf{i}\,\frac{\partial}{\partial x} + \mathbf{j}\,\frac{\partial}{\partial y} + \mathbf{k}\,\frac{\partial}{\partial z}, \tag{2.62}$$

where he used "d" instead of "∂". This operator, ∇, is variously known as grad, del, nabla, and atled. It has frequent use in fluid dynamics:

a)   Multiplied by a scalar such as "s", it becomes

$$\nabla s = \mathbf{i}\,\frac{\partial s}{\partial x} + \mathbf{j}\,\frac{\partial s}{\partial y} + \mathbf{k}\,\frac{\partial s}{\partial z}, \tag{2.63}$$

which describes the *gradient* of s as the sum of the individual gradients in each of the three orthogonal Cartesian coordinate directions.

b)   If the gradient of a scalar variable "s" is dot multiplied by a distance vector, $d\mathbf{r}$,

$$\nabla s \cdot d\mathbf{r} = \frac{\partial s}{\partial x}\,dx + \frac{\partial s}{\partial y}\,dy + \frac{\partial s}{\partial z}\,dz = ds, \tag{2.64}$$

it produces the total differential, δs over the distance δr (note that time is not included). As a practical example, if we divide by dt, replace s by T and add $\partial T / \partial t$ to both sides, we may rewrite Equation (2.64) as

$$\frac{\partial T}{\partial t} + \nabla T \cdot \frac{d\mathbf{r}}{dt} = \frac{\partial T}{\partial t} + \frac{\partial s}{\partial x}\frac{dx}{dt} + \frac{\partial T}{\partial y}\frac{dy}{dt} + \frac{\partial T}{\partial z}\frac{dz}{dt} = \frac{dT}{dt}. \tag{2.65}$$

Furthermore, since $d\mathbf{r}/dt$ may also be replaced by $\mathbf{V}$,

$$\frac{dT}{dt} = \frac{\partial T}{\partial t} + \nabla T \cdot \mathbf{V}. \tag{2.66}$$

This may be interpreted physically as "the rate of change of T, for example, temperature, following a parcel, is equal to the rate of change of T at an observation point in the coordinate system plus the movement (advection, and convection) of the T field." The equation is known as

Euler's relation. Often Equation (2.66) is turned around so that temperature change over time measured at a station $(\partial T/\partial t)$ is equal to the time temperature changes in a parcel due to heating and cooling processes $(dT/dt = $ radiation, turbulence, adiabatic changes) minus any transfers (advection, and convection).

c) If grad is dot multiplied by a velocity vector

$$\nabla \cdot \mathbf{V} = \frac{\partial u}{\partial x} + \frac{\partial v}{\partial y} + \frac{\partial w}{\partial z} \qquad (2.67)$$

it becomes the *divergence of the vector* $\mathbf{V}$. Applied to fluid flow it states that an increase in speed in the direction of motion must be accompanied by a fractional decrease in the density (see Equation 7.3).

d) When grad is cross multiplied by a vector it produces the *curl* of a vector

$$\nabla \times \mathbf{V} = \begin{vmatrix} \mathbf{i} & \mathbf{j} & \mathbf{k} \\ \dfrac{\partial}{\partial x} & \dfrac{\partial}{\partial y} & \dfrac{\partial}{\partial z} \\ u & v & w \end{vmatrix} \qquad (2.68)$$

$$= \mathbf{i}\left(\frac{\partial w}{\partial y} - \frac{\partial v}{\partial y}\right) + \mathbf{i}\left(\frac{\partial u}{\partial z} - \frac{\partial w}{\partial x}\right) + \mathbf{k}\left(\frac{\partial v}{\partial x} - \frac{\partial u}{\partial y}\right)$$

$$(2.69)$$

The last term, the rotation in the horizontal plane, is the familiar *vorticity* (see section 7.10).

e) If grad is dot multiplied by itself, it becomes ∇^2, which is a scalar and is known as the *Laplacian*, although it may have originated with Euler (Youschkevitch, 1981),

$$\nabla \cdot \nabla s = \nabla^2 s = \frac{\partial^2 s}{\partial x^2} + \frac{\partial^2 s}{\partial y^2} + \frac{\partial^2 s}{\partial z^2}. \qquad (2.70)$$

When this is equal to zero it is known as *Laplace's equation* and when equal to $(1/\kappa)(\partial s/\partial t)$ the *diffusion equation*. If s is temperature and κ is the thermal diffusivity this is the *equation of heat conduction*. When the Laplacian of s is equal to $(1/c^2)(\partial^2 s/\partial t^2)$ it becomes the *wave equation*.

Not only do we remember Laplace for his differential equations we recognize his impact on a wide range of fields. Indeed, he was one of the most influential scientists in history. Pierre-Simon, Marquis de Laplace (1749–1827), born in Normandy, initially concentrated on theology at the University of Caen. At 19, however, with an aptitude and interest in mathematics, he moved to Paris with a letter of recommendation to d'Alembert. As a professor at the École

Militare Laplace soon gained recognition in calculus, astronomy, and probability. He published major treatises, *Méchanique célestre* (Laplace, 1825) in five volumes between 1799 and 1825 and *Théorie analytique des probabilitiés* in 1812. His ideas on science were spread through his service on various government committees beginning in the 1780s. Among other projects he was involved in the establishment of the famous *École Polytechnique* that was home for numerous French scientists (see École in the index) at the beginning of the nineteenth century (Grattan-Guinness, 1981).

This Parisian engineering school had its beginnings in 1794 as a Central School of Public Works under Lazare Carnot and Gaspard Monge. Directed by the Ministry of Defense it took its present name in 1795. In 1802 it absorbed the state military school and actually became a military school in 1804 under Napoleon. It was moved to Palaiseau in 1976.

2.6.5 Tensors

Vector mathematics became much more formalized with the work of Oliver Heaviside (1850–1925), an English physicist, and with J. Willard Gibbs (1839–1903) a professor of mathematical physics at Yale from 1871 until his death. Gibbs lectured annually on vector analysis and Edwin Bidwell Wilson, an instructor in mathematics at Yale, wrote a textbook founded upon those lectures and the publications of Heaviside (Wilson, 1901).

From this brief introduction it should be evident that vector notation is one way of handling variables that have the attributes of magnitude and direction. Also, although it may appear complex to the uninitiated, it often significantly reduces the number of terms in an equation thus rendering the mathematics more concise.

In fact, vector mathematics is a subset of tensor mathematics. Scalars, having one component, are said to be tensors of zero order and vectors, having three components, are tensors of first order. Tensors of second order have nine components. The latter enter fluid dynamics because stresses on a plane face exist in both the normal and parallel directions (i.e. the pressure gradient is normal, friction is parallel). While workers in fluid dynamics typically invoke tensors, atmospheric scientists usually do not. For a discussion of tensors see, for example, Jeffreys (1931), Aris (1962), Defrise (1964), and Brown (1991).

2.7 The Exponential and Complex Numbers

2.7.1 The Number e

A number which has important use in mathematics is labeled e, the base of Naperian logarithms. The Scot, John Napier (1550–1617), did not discover this number although he was close to it when he produced the first tables, which took him twenty years (1594–1614), and called them *logarithms*. Maor (1994) suggests that Napier arrived at the idea of logarithms, thereby making multiplication and division easier by replacing them by addition and subtraction, from the well known trigonometric relationships such as

$$\sin(A) \times \sin(B) = \frac{1}{2}[\cos(A - B) - \cos(A + B)],$$

but there is no evidence for that.

The origin of e may be found in the concepts of compound interest that date at least from 1700 BC in Mesopotamia. The precise definition had to await the introduction of decimals (sixteenth century), logarithms (Napier), the limit, and the general solution of the binomial (Newton).

Ultimately e was defined by

$$e = \lim_{n \to \infty} \left[1 + \frac{1}{n} \right]^n. \tag{2.71}$$

But the binomial expansion gives

$$\left[1 + \frac{1}{n} \right]^n = 1 + n\frac{1}{n} + \frac{n(n-1)}{2!}\left(\frac{1}{n^2}\right) + \frac{n(n-1)(n-2)}{3!}\left(\frac{1}{n^3}\right) + \cdots$$

$$= 1 + 1 + \frac{1}{2!}\left(1 - \frac{1}{n}\right) + \frac{1}{3!}\left(1 - \frac{1}{n}\right)\left(1 - \frac{2}{n}\right)$$

$$+ \frac{1}{4!}\left(1 - \frac{1}{n}\right)\left(1 - \frac{2}{n}\right) + \left(1 - \frac{3}{n}\right) + \cdots,$$

where the symbol "!" stands for *factorial*, i.e. $n! = n \times (n-1) \times (n-2) \times \cdots \times 2 \times 1$. Therefore

$$e = 1 + 1 + \frac{1}{2!} + \frac{1}{3!} + \frac{1}{4!} + \frac{1}{5!} \cdots,$$

which, to eight significant digits is

$$e = 2.7182818. \tag{2.72}$$

Maor (1994) and Brothers and Knox (1998) have given other expansions.

The first person to use e was Euler for the base of natural logarithms in 1727 although the publication date was 1736 (Cajori, 1929). Logarithms are defined by

$$N = b^L, \tag{2.73}$$

where N is a number, b is any base, and L is the logarithm, so

$$L = \log_b(N). \tag{2.74}$$

When $b = e$ the logarithm is called the *natural*, or *Naperian*, logarithm and Equation (2.74) is written $L = \log_e(N) = \ln(N)$.

The natural logarithm is seen most frequently in meteorology in the integration of $1/p$, i.e.

$$\int \frac{dp}{p} = \ln p + c, \tag{2.75}$$

but it has wider use in calculus. For example,

$$\frac{d(\log_{10} t)}{dt} = \frac{1}{t} \log_{10} e, \tag{2.76}$$

$$\frac{d(u^t)}{dt} = u^t \ln u, \tag{2.77}$$

$$\frac{d(e^{ut})}{dt} = u e^{ut}, \tag{2.78}$$

and

$$\int e^{ut} \, dt = \frac{1}{u} e^{ut} + const. \tag{2.79}$$

Equation (2.78) shows that the function e^t equals its own derivative. Therefore it is the slope of the graph of $y = e^x$ at $x = 1$. It is also the area under the same graph from $x = -\infty$ to $x = 1$. For more see Maor (1994).

2.7.2 The Imaginary Number

The term *imaginary number*, i, is applied to the root of $x^2 + 1 = 0$, and is often expressed as $i = \sqrt{-1}$.

The development of algebra based on the square root of negative numbers has been attributed to Raphael Bombelli (1526–72). Most of his professional career was spent as an engineer-architect in the service of a Roman nobleman Alessandro Rufini. He worked on the reclamation of marshes. Because he found existing texts unclear he wrote his own *Algebra* during the years 1557 to 1560. Only part was published by the time of his death and the whole text did not appear until 1929. Nevertheless Bombelli's work was widely acknowledged in mathematical circles and Leibniz is reported to have stated that Bombelli was an "outstanding master of analytical art" (Jayawardene, 1981).

One of the problems that Bombelli investigated was the solution for cubic equations. He showed that there was "another kind of cube root." He called the square roots of a negative quantity *più di meno* and *meno di meno* for the positive and negative roots respectively. Euler used the letter i to represent the imaginary number in 1777 although, as was his normal practice, he delayed publishing this designation for several years until 1794 (Youschkevitch, 1981).

2.7.3 Complex Numbers

Many mathematicians of the seventeenth and eighteenth centuries, such as Descartes, Newton, and Euler, used the imaginary but it was H. Kühn who first gave it a geometrical representation in 1750. Later Jean Robert Argand (1768–1822) extended this graphical approach. Very little is known about the man except that he was a Parisian bookkeeper who published his work on complex numbers in 1806. There is some suggestion that he was aware of the works of Wallis, de Moivre, and Euler. He did have contact with Legendre.

Complex numbers are number pairs composed of a real and imaginary part. The latter is distinguished by having $i = \sqrt{-1}$ as a factor, i.e. the complex number c is given by $c = a + (\sqrt{-1})b$. The use of the letter i to represent $\sqrt{-1}$ was popularized by Gauss but, because in some fields i represents "current," it is often replaced by the letter "j." The complex number pair therefore is written

$$c = a + ib. \tag{2.80}$$

What Argand did was to place the real part of the complex number on the abscissa and the imaginary part on the ordinate. Then the complex number itself was represented by a point in the complex plane on this graph, now known as an *Argand Diagram*. The point could be thought of as a vector with one component on the real axis and the other on the imaginary axis. Indeed, Argand used vectors with the overbar notation, e.g. $\overline{KA}$. He also showed that complex roots come in pairs, i.e. one being the *conjugate* of the other. The conjugate is obtained by changing the sign of i. It is also represented by the exponent "*," e.g. c^*,

$$c^* = a - ib. \tag{2.81}$$

Thus

$$c \times c^* = (a + ib)(a - ib) = a^2 + b^2. \tag{2.82}$$

The square root of this quantity is variously known as the *modulus* or *absolute value* of c, and may be represented with vertical lines, i.e.

$$|c| = \sqrt{a^2 + b^2}. \tag{2.83}$$

The angle between the vector from the origin to the point (a, b) and the abscissa is given by

$$\Phi = \arctan\left(\frac{b}{a}\right). \tag{2.84}$$

2.7.4 Cosines and Sines as Complex Numbers

Maclaurin's Series

A variable $x[t]$ may be written as a polynomial in t,

$$x[t] = C_0 t^0 + C_1 t + C_2 t^2 + C_3 t^3 + \cdots + C_n t^n + \cdots, \qquad (2.85)$$

Repeated differentiation of Equation (2.85) gives

$$x'[t] = 1C_1 + 2C_2 t + 3C_3 t^2 + \cdots$$
$$x''[t] = 2 \cdot 1 \cdot C_2 + 3 \cdot 2 \cdot C_3 t + \cdots$$
$$x'''[t] = 3 \cdot 2 \cdot 1 \cdot C_3 + \cdots.$$

Set $t = 0$ and we have

$$x[0] = C_0$$
$$x'[0] = C_1 1!$$
$$x''[0] = C_2 2!$$
$$x'''[0] = C_3 3!$$
$$\vdots$$
$$x^n[0] = C_n n!$$

Substitute into Equation (2.85) and we have

$$x[t] = x[0] + \frac{x'[0]t}{1!} + \frac{x''[0]t^2}{2!} + \frac{x'''[0]t^3}{3!} + \cdots + \frac{x^n[0]t^n}{n!} + \cdots. \qquad (2.86)$$

This is known as *Maclaurin's Series*. Colin Maclaurin (1698–1746), brought up in a religious Scottish family, entered the University of Glasgow in 1709 to study divinity but within a year had switched to mathematics. He was appointed to a mathematics professorship at Aberdeen while still in his teens. He traveled to France and to London where he became friends with Newton. The latter was influential in getting Maclaurin a chair in 1725 at Edinburgh where much of his mathematical work was done. Besides the series presented above, Maclaurin is recognized for his systematic presentation of Newton's methods in his 1742 publication of *Treatise of Fluxions*.

Application

The Maclaurin expansion of e^t and e^{-t}, with the use of Equation (2.78) for $u = 1$, may be written

$$e^t = 1 + t + \frac{t^2}{2!} + \frac{t^3}{3!} + \frac{t^4}{4!} + \cdots + \frac{t^n}{n!} + \cdots \tag{2.87}$$

and

$$e^{-t} = 1 - t + \frac{t^2}{2!} - \frac{t^3}{3!} + \frac{t^4}{4!} - \cdots - \frac{t^n}{n!} + \cdots . \tag{2.88}$$

where in this example n is odd. Similarly applied to e^{it} and e^{-it} we obtain

$$e^{it} = \left(1 - \frac{t^2}{2!} + \frac{t^4}{4!} + \cdots\right) + i\left(t - \frac{t^3}{3!} + \frac{t^5}{5!} + \cdots\right). \tag{2.89}$$

and

$$e^{-it} = \left(1 - \frac{t^2}{2!} + \frac{t^4}{4!} + \cdots\right) - i\left(t - \frac{t^3}{3!} + \frac{t^5}{5!} + \cdots\right). \tag{2.90}$$

Also, since the differentiation of $+\cos(t)$ is $-\sin(t)$ and of $+\sin(t)$ is $+\cos(t)$ the expansions of these two trigonometric functions are

$$\cos(t) = 1 - \frac{t^2}{2!} + \frac{t^4}{4!} - \frac{t^6}{6!} + \cdots \tag{2.91}$$

and

$$\sin(t) = t - \frac{t^3}{3!} + \frac{t^5}{5!} - \frac{t^7}{7!} + \cdots . \tag{2.92}$$

From Equations (2.89), (2.90), (2.91) and (2.92) we have

$$e^{it} = \cos(t) + i\sin(t) \tag{2.93}$$

and

$$e^{-it} = \cos(t) - i\sin(t) \tag{2.94}$$

A form of these was first deduced by Abraham de Moivre (1667–1754) who is known for his work in probability, having laid the groundwork for the standard deviation and the normal curve of Laplace and Gauss (see chapter 3). However, R. Cotes in 1716 and Euler in 1743 also gave formulations. Euler can be credited too as the first to arrive at the equality linking the mathematical numbers i, π, and e: $\ln(-1) = \pi i$, or

$$e^{\pi i} = -1 \tag{2.95}$$

(Youschkevitch, 1981).

2.8 Finite Differences

2.8.1 Development

Calculus allows us to investigate continuous processes in the real world such as the flow of fluids. In practice, however, we still have only finite observations of our phenomena whether it be in time or space. We still rely on atmospheric soundings made at a limited number of stations and even high resolution remotely sensed satellite data of the earth are composed of individual pieces called pixels. We may attempt to fit mathematical functions, such as Fourier series, to these data and then apply our differential equations directly to them. However, even though our theory is set down in differential equations, many of our calculations must be performed with finite differences i.e. $x_2 - x_1 = \Delta x$, a notation introduced by Euler in 1755 (Youschkevitch, 1981). This means we must go from

$$\frac{\partial y}{\partial x} \text{ to } \frac{\Delta y}{\Delta x}.$$

However, as we have seen, these are not equal unless Δx approaches zero. A solution is made possible by the series developed by the English mathematician and supporter of Newton, Brook Taylor (1685–1731). He attended St. John's College, Cambridge, in 1701 and obtained an LLB in 1709 and an LLD in 1714. He became a fellow of the Royal Society in 1712 and was immediately selected with A. de Moivre and F. Ashton to serve on the committee investigating the dispute between Newton and Leibniz. Partly as a result of that association, his mathematical work incurred the criticism of Johann Bernoulli and Leibniz. In 1715 he showed that

$$f(y + x) = f(y) + xf'(y) + \frac{x^2}{2!}f''(y) + \frac{x^3}{3!}f'''(y) \cdots , \qquad (2.96)$$

which is known as the Taylor series (Feigenbaum, 1981). Each prime represents a differentiation with respect to x.

In 1899 Sheppard made use of the Taylor series in extending the accuracy of mathematical tables (Sheppard, 1899). Then Lewis Fry Richardson in 1910, in response to suggestions by Karl Pearson, showed how such finite differences might be used to solve equations like the diffusion Equation (2.70), e.g.

$$\frac{\partial^2 \chi}{\partial s^2} = \frac{1}{R}\frac{\partial \chi}{\partial q}. \qquad (2.97)$$

in association with his investigation into the stresses in a masonry dam (Richardson, 1910). As we shall see in section (9.2.1) Richardson subsequently applied these techniques in his famous attempt at the prediction of atmospheric pressure (Richardson, 1922).

2.8.2 Example

To illustrate the technique we shall consider the partial derivative of temperature, T, as a function of time, t. With the substitution of Δt for x and $T(t)$ for y in Equation (2.96)

$$T(t + \Delta t) = T(t) + \Delta t \, \frac{\partial T(t)}{\partial t} + \frac{(\Delta t)^2}{2!} \frac{\partial^2 T(t)}{\partial t^2} + \frac{(\Delta t)^3}{3!} \frac{\partial^3 T(t)}{\partial t^3} + \cdots,$$

$$(2.98)$$

and

$$T(t - \Delta t) = T(t) - \Delta t \, \frac{\partial T(t)}{\partial t} + \frac{(\Delta t)^2}{2!} \frac{\partial^2 T(t)}{\partial t^2} - \frac{(\Delta t)^3}{3!} \frac{\partial^3 T(t)}{\partial t^3} + \cdots,$$

$$(2.99)$$

Each of the above two equations provides one estimate of $\partial T(t)/\partial t$. From the first

$$
\begin{aligned}
\frac{\partial T(t)}{\partial t} &= \frac{T(t + \Delta t) - T(t)}{\Delta t} - \frac{\Delta t}{2!} \frac{\partial^2 T(t)}{\partial t^2} - \frac{(\Delta t)^2}{3!} \frac{\partial^3 T(t)}{\partial t^3} - \cdots \\
&= \frac{T(t + \Delta t) - T(t)}{\Delta t} + O(\Delta t).
\end{aligned}
$$

$$(2.100)$$

This says that the direct substitution of a finite difference for a partial derivative involves a first order error (Δt has the exponent 1), the error being called a truncation error and being represented by $O(\Delta t)$. A more accurate form may be obtained by subtracting Equation (2.99) from Equation (2.98).

$$
\begin{aligned}
\frac{\partial T(t)}{\partial t} &= \frac{T(t + \Delta t) - T(t - \Delta t)}{2 \Delta t} - \frac{(\Delta t)^2}{3!} \frac{\partial^3 T(t)}{\partial t^3} - \cdots \\
&= \frac{T(t + \Delta t) - T(t - \Delta t)}{2 \Delta t} + O(\Delta t)^2,
\end{aligned}
$$

$$(2.101)$$

which involves a smaller second order error (Δt has the exponent 2).

So as to illustrate the technique we shall consider the diffusion equation as applied to heat conduction in one dimension,

$$\frac{\partial T}{\partial t} = \frac{1}{C} \frac{\partial}{\partial z} \left(k \frac{\partial T}{\partial z} \right),$$

$$(2.102)$$

where T is temperature in K, z is length in m, k is thermal conductivity in $J\,s^{-1}\,m^{-1}\,K^{-1}$ and C is the heat capacity in $J\,m^{-3}\,K^{-1}$. If the thermal conductivity is independent of distance then the equation may be written

$$\frac{\partial T}{\partial t} = \kappa \frac{\partial^2 T}{\partial z^2}, \tag{2.103}$$

where $\kappa = k/C$ is the thermal diffusivity in $m^2\,s^{-1}$.

The early investigator of heat diffusion with partial differential equations was Jean Baptiste Joseph Fourier (1768–1830). Orphaned at nine he was placed in Auxerre's military school where he became excited by mathematics. He was temporarily jailed twice by different factions in the Revolution and worked for a short time for Lagrange and Monge at the École Polytechnique. At the end of the century he served 3 years as a diplomat for Napoleon in Egypt and then as prefect in Grenoble. When Napoleon fell in 1815 Fourier resigned. His mathematical work was already recognized and in 1822 he was elected *secretaire perpétuel* of the Académie des Sciences. Much of the last years of his life he was confined to his home due to an illness, possibly myxedema, a slowing down of the body mechanisms, that he had contracted in Egypt. It was in that country where he started his heat diffusion research that he first presented in 1807. Despite support from Laplace, Monge and Lacroix, this work was strongly opposed by Lagrange because of Fourier's use of trigonometric series (see 3.4.2 below). As a result, the work was not fully published until 1822 (Fourier, 1822).

More recently the analytical aspects of heat diffusion have been studied intensively by Carslaw and Jaeger (1959). Also, an easily understood derivation of the equation for soil is given by Sellers (1965).

Equation (2.103) written in simple finite differences for a particular layer in the soil is

$$T[z, t + \Delta t] \approx \lambda T[z + \Delta z, t] + (1 - 2\lambda)T[z, t] + \lambda T[z - \Delta z, t], \tag{2.104}$$

where $\lambda = \kappa\,\Delta t/(\Delta z)^2$. Here the left hand side represents the temperature in layer z at a time Δt in the future from the time t. The three terms on the right hand side depend on the temperature at time t and in the layer below, $z + \Delta z$; in that layer, z; and the layer above, $z - \Delta z$; respectively. There is also the coefficient λ. In order to solve this equation, or set of equations if several layers are involved, the temperature of the top and bottom layers must be given. In this context these inputs are known as *boundary conditions*. Clearly for a reasonable set of results for an intermediate layer these boundary conditions must be realistic. In addition, the calculation must begin with some magnitudes of temperature. These are known as *initial conditions*. In contrast to boundary conditions, initial conditions are often less critical for the results. For example, the initial temperatures at all layers could all be set equal to some constant such as the mean, as has be done in the illustration below.

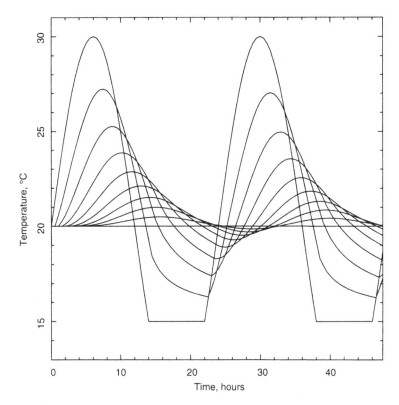

Figure 2.13 Numerical solution of the diffusion equation applied to 10 layers

If unrealistic initial conditions are introduced in such cases it would just take longer for the system to come to equilibrium and produce realistic results. In certain situations with a different model the initial conditions may control the final equilibrium outcome (see section 9.1.1).

To illustrate the simple diffusion of heat model the bottom layer temperature has been taken to be a constant. This is reasonable since not too far down in the soil we typically find a level where the temperature remains constant throughout a diurnal period. On a clear day the temperature of the top layer is controlled by radiation, mainly the sun. Heat flows also occur by turbulence (latent and sensible) but they are highly correlated with the solar flux. As a gross approximation we may use a sinusoidal curve to represent solar absorption while the sun is above the horizon. At night it is a constant, zero. Then the net outgoing radiation is in the longer wavelengths but it is relatively constant and does not approximate a sinusoid. Hence to approximate the top layer temperature a sinusoid has been chopped off and replace by a constant (see figure 2.13).

The results displayed in that figure are for ten layers for two days. They are essentially the same as might be obtained from an analytical solution which is available for such a system. The initial temperatures were all set at the bottom layer mean. Therefore the temperatures all diverge from there and take several time steps to adjust to the *forcing* by the surface boundary conditions. It is evident that the model of this system responds quickly and essentially comes to equilibrium after one forcing cycle. Depending upon the model it may take many cycles to approach equilibrium, if ever.

In principal, as we shall see, the current method for forecasting and ultimately for explaining climate follows this approach. Even for the soil heat flow component the model just described is too simplistic. For forecasting, the initial conditions are usually taken from the current fields modified for current observations. Because of the numerous feedback mechanisms climate models are usually run for several simulation years to allow all variables to come to equilibrium before any experimental data are retained for analysis.

The simple heat diffusion model may also be used to illustrate one of the problems that arise from finite difference solutions. The coefficient λ must be less than 0.5 for the calculations to produce a stable result. λ is composed of physical "constants" plus the ratio of the temporal and spatial increments. The constants are set by the physics so we may control only the increments: given one increment, say space Δz, the increment of the other, time Δt, has an upper limit. If this constraint is not observed the result will be unstable. In figure 2.13 λ is 0.44. For the solution for temperature at the third layer shown as a dashed line in figure 2.14 it is 0.53. As is evident, the predicted temperature is quickly diverging from its expected magnitudes.

Current forecast models which use horizontal spacings of the order of 100 km need to use time steps of a few minutes. That is one of the reasons why weather forecast models need large and fast computers. The forecast for a day ahead needs in the order of 10^3 intervening forecasts. However, not all variables in a large model have to be calculated that frequently.

2.9 Comment

This chapter has introduced a restricted set of mathematical relationships. Some are quite elementary, as in section 2.1. Their selection was based upon their use in atmospheric analysis. Others might have been included and future developments in meteorology will dictate that. The content has therefore been driven by the application. Hence we may describe the review as one into *applied mathematics*, essentially the only type of mathematics before about 550 BC. It was Pythagoras (ca. 582–507 BC) who is usually identified as the one who moved mathematics into the realm of the study of the field for its own sake, for its own beauty, without regard to its applicability to the real

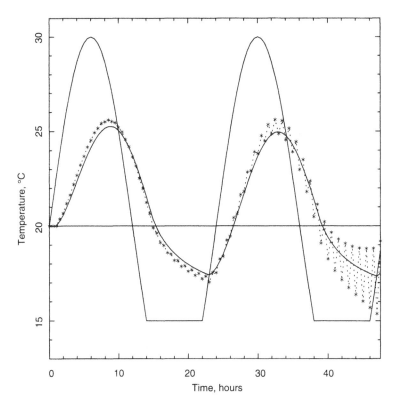

Figure 2.14 Numerical solution of the diffusion equation applied to 10 layers. The boundary conditions (shown) are the same as in figure 2.13. A continuous line represents the temperature of third layer for $\lambda = 0.44$ and the dashed line the equivalent for $\lambda = 0.53$

world. It established procedures whereby mathematical relationships could be rigorously proven. This approach has become known as *pure mathematics*. Simultaneously, advances in mathematics could still be driven by the need to solve practical problems (e.g. calculus for the problems that Newton was investigating). Now, however, any new technique has to be put on a more fundamental mathematical footing. This tends to make it more abstract. Courses in mathematics that are prerequisites for fields such as atmospheric science are often more "pure" than "applied," although most teachers try to select examples that are relevant to the applied fields. Sometimes, because the content of these courses appear to have only limited direct application, students tend to see them in a negative light. This is unfortunate. Topics, which may be considered esoteric in one period, may have great value in applied fields in a later period.

The point to be made here is that anyone, who intends to pursue scientific work that involves mathematics (virtually all), should take as many pure mathematics courses as they are able. Basic grounding and breadth are equally important. Without a good understanding of the logic involved, mistakes in application are likely and, without breadth, the researcher may not recognize that solutions exist that may lead to early scientific breakthroughs.

This chapter, like the others in this book, is not meant to be a replacement for prerequisite courses. It is meant as a review. If it is the first exposure to some of the ideas, then hopefully it will encourage the reader to search out the more formal exposition in the literature and/or in courses.

Chapter 3

Statistics

3.1 Data

Climatology and meteorology are fields generating enormous amounts of data. Not only are there vast numbers of observations, there are also computer generated fields from interpolation and extrapolation. There are about 1,200 stations reporting standard variables recorded from instruments carried aloft twice a day by balloons (figure 3.1), and about seven times that number of synoptic stations recording conditions (some continuously) from the surface (figure 3.2). In addition, many more locations have daily records of precipitation and maximum and minimum temperatures as well as irregular ship and aircraft data. As is evident from the maps these stations are by no means well spaced over the globe. Most occur in the developed countries (see appendix D). The ocean areas covering 71 percent of the surface are not well represented although more buoy data have become available recently. The poor distribution of observations has forced many researchers to rely on the interpolated meteorological fields that have been generated by numerical models. The procedure takes into account the time dimension as well as the physical relationships so thus provides better estimates than a simple spatial interpolation could supply. In addition, more and more remote sensing products are becoming available. Current satellite images with resolutions of approximately 1 km and radar mosaics are generating gibibytes (see appendix A) of data daily. Somehow, all this information has to be arranged into a form that can be comprehended. Analyses of observations have occupied many many researchers but it has usually been performed on very restricted data sets. This is understandable

Upper air stations 1998

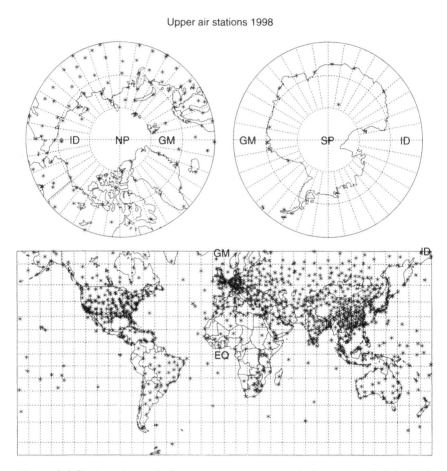

Figure 3.1 Stations that typically report one or two vertical soundings daily in 1998

since the problems, for which the analyses have usually been applied, have had limited generality; the data are extremely variable in their reliability; and, as indicated, they tend to be spatially concentrated. In the last two decades of the twentieth century the situation changed and a more global view is possible.

Various approaches to assimilation have been taken. One has been the use of case studies in which "typical" situations have been selected. The resulting analyses have often produced a stereotype model of the conditions of interest. For example, the Norwegian extra-tropical cyclone model developed during

Surface synoptic stations 1998

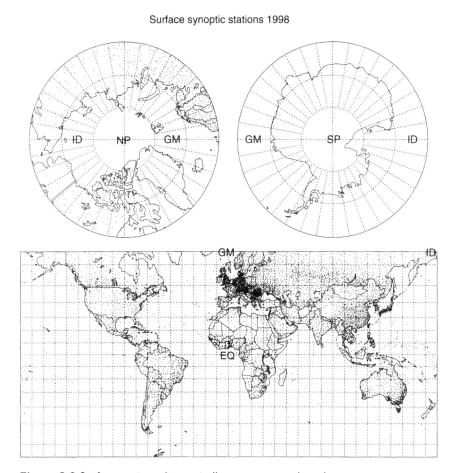

Figure 3.2 Surface stations that typically report every three hours

the second decade of the twentieth century is of that type. A slightly different method is that of *compositing* which has provided a way of generalizating systems. The tropical cyclone has been reduced to a type of averaged model by such calculations (Gray, 1978).

A more general approach is the use of statistics. However, as argued by Brown et al. (1999) formal training in statistics and probability for atmospheric students, especially in the United States, remains minimal. Only a brief and selective introduction to the subject is given below, and like the other topics reviewed in this book, it cannot replace a full sequence of courses.

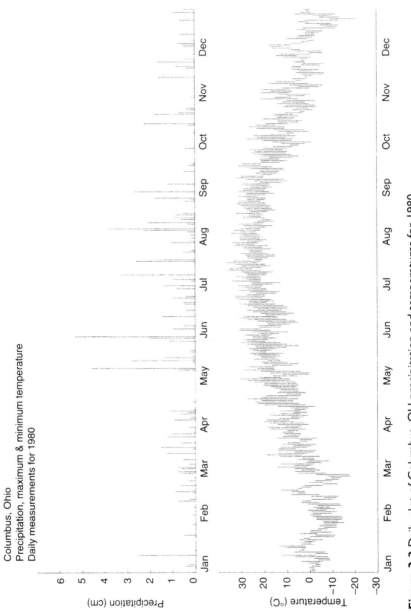

Figure 3.3 Daily plot of Columbus, OH precipitation and temperatures for 1980

3.2 One Variable Descriptive Statistics

A typical example of climatological data is displayed in figure 3.3. One year of daily precipitation and minimum to maximum temperatures are plotted (original observations in inches and degrees fahrenheit have been converted to millimeters and celsius). Daily mean temperatures are obtained by adding the minimum and maximum and dividing by two. Since 1980 was a leap year 366 mean temperatures are produced. The plots are typical of mid-latitude locations. The precipitation appears quite irregular both in amount and in interval between occurrence with highest individual days occurring in summer. The temperatures, display significant fluctuations over a few days especially in winter and a clear annual oscillation. The question is, "How might these numbers be summarized?" As a first step we could count the number of observations that fall within predetermined non-overlapping classes, thus producing a frequency distribution. The frequencies themselves indicated how often a given class has been observed. In other words, they are indicators of the probability of occurrence of the different classes. The precipitation frequencies (figure 3.4) reveal a steep decline from high values at low amounts to low values spread out in a long tail towards high magnitudes. "Trace amounts" which are recorded on days when the precipitation is less than 0.01 inches are included in the zero class. The shape of this distribution is typical for precipitation. The mean temperature frequencies (figure 3.5) display a rather different form with relatively steep tails on either end of a group of higher but fluctuating counts. Most meteorological data produce frequency distributions of the temperature type.

So, we have reduced the 366 raw observations to about 50 frequencies in each case. As a result of our analysis we have also found out how often certain events occurred, e.g. on 234 of the days (64 percent) in 1980 this station recorded only a trace or no precipitation. The 50 new numbers, however, are still too many for us really to assimilate and we need to summarize even more. Especially for temperature-type distributions it is useful to indicate where the data are centered in the continuum of temperatures. For such a purpose three different measures of central tendency have been identified. The oldest is the *mean*, having been used before Pythagoras. It is the sum of the observations divided by the number of observations. For the precipitation and temperature data in figure 3.3 the means are 2.60 mm and 10.7°C respectively. The second is the *median*, the middle of the frequency distribution in terms of the number of observations, i.e. half above and half below. It was introduced as a concept in statistics by Galton in 1869 and by name in 1883 (Galton, 1883). From the frequency counts they fall in the 0 mm and 11°C centered classes respectively. The third is the *mode*, introduced by Pearson (1894), the most frequent, the most probable value. For the temperature data the class having the highest frequency is 23°C but other classes such as −1°C and 19°C also contain large counts: these data appear to have several modes.

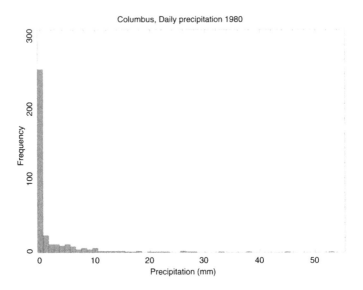

Figure 3.4 Histogram of precipitation data displayed in figure 3.3

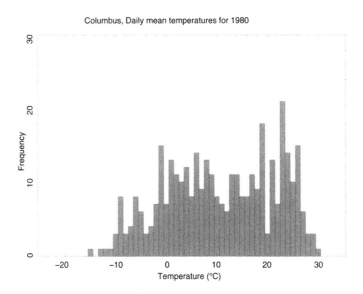

Figure 3.5 Histogram of temperature data displayed in figure 3.3

Of the various measures of location the mean is the most widely used. Indeed, climate has many times and misleadingly been defined as average weather. However, by itself the mean has limited value. For example, we may observe that both Columbus, US, and Southampton, UK, have approximately the same annual mean temperatures near 11°C and yet we know that the temperatures of these two cities are rather different. Just one set of published extreme minima to maxima of $-29°$ to $41°$ and $-12°$ to 30°C respectively show that. Hence it is important to provide some measure of dispersion of the data such as the mean sum of the magnitudes of deviations from the mean. Since the sum of the deviations from the mean is zero it is necessary to ignore the sign. The mean sum of squares of deviation from the mean does not have this problem although it does emphasize observations which lie a long way from the mean. This measure was given the name *variance* by Fisher (1920). The variance of the 1980 precipitation data is 40.38 mm^2 and of temperature is 115.24°C^2. As we shall see later, by itself, variance is only really useful if we know more about the form of the frequency distribution.

We should now stand back and consider in more depth what we are doing. We began by attempting to summarize enormous amounts of data. The assumption was that we may make valuable generalizations about what has been observed. A more exciting and perhaps underlying motive is to say something about what we can expect in the future. These are related but different activities. To elaborate we need to review some simple statistical theory.

In statistics, observations are considered to be taken from an *ensemble*. That is, when we make estimates of cloud amount we could make many observations at the same instant. For example, we might have 20 observers all in the same place making the same observation. The numbers that statisticians calculate, such as the mean, are then made using that ensemble of observations. Since this is only an estimate based on that set, or *sample*, of observations taken from the total theoretical *population* it is called a *statistic*. It is used to represent the "true" or theoretically ideal number, which is given the name *parameter*. In order to distinguish statistics from parameters they are either assigned different symbols or the statistic is represented by a hat, "ˆ," over the parameter symbol. For example, the Greek letter μ with the subscript "1" is usually chosen for the population mean and $\hat{\mu}_1$ or $\overline{X}$ for the sample mean. Similarly μ_2 or σ^2 are typically used for the population variance and $\hat{\mu}_2$ or $\hat{\sigma}^2$ or s^2 for the sample variance. Obviously we would like our statistic to be as close to the parameter as possible and we would also like to have a measure of its reliability.

First, the observations upon which the statistic is based should be accurate and representative. Observations may be made with varying degrees of precision. For example, estimates of cloud amount by observing the sky and selecting a number between 0 and 10 that appear to the observer to be that proportion of the sky that is covered by cloud. At other times we may measure the temperature of the air with known accuracy to the nearest 1/10 of a degree. However, if the thermometer were not located correctly with shielding from the sun, subject to appropriate air flow, etc., the precise reading may not represent the true air temperature as well as the subjective estimate of cloud represented the true cloud amount. Problems of data accuracy and precision can be the subject of a text by itself. It is important to realize that just because a measurement has been recorded and published it may not accurately represent the true variable for that location at that instant. Similarly the measurement may not be representative of the region in which it was made. The latter is a problem in sampling. The location of the measurement point may be atypical or the element may be so variable that one measurement location cannot characterize that element. Again a large literature exists on spatial and temporal sampling. In meteorology temporal sampling is dependent upon the choice of δt, the constant interval between observations. Spatial sampling is a far more significant problem. The location of stations is a function of the variation in terrain, especially land versus sea, as well as political and economic factors. Often, as in the case of numerical modeling, data from irregularly spaced sites are interpolated to some form of regular grid. Then some guidance is usually provided by the forecast field for that particular time stamp.

Secondly, the *estimator*, the procedure, or equation selected for calculating the statistic, influences how well the result might match the parameter. There is really only one estimator used for the mean, which for discrete observations is given by

$$\hat{\mu}_1 = \frac{1}{n} \sum_{i=0}^{n-1} x[i]. \tag{3.1}$$

For the variance, different estimators are available depending upon the divisor, i.e. we may use n, $n - 1$, etc. For a divisor of $n - 1$, the variance may be estimated with

$$\hat{\mu}_2 = \frac{1}{(n-1)} \sum_{i=0}^{n-1} (x[i] - \hat{\mu}_1)^2 = \frac{1}{(n-1)} \left[\sum_{i=0}^{n-1} x^2[i] - \frac{\left(\sum_{i=0}^{n-1} x[i] \right)^2}{n} \right] \tag{3.2}$$

Obviously, such estimators that modify n by one produce significantly different estimates only when n is small. Usually the larger n the better. The quality of an estimator can be judged by applying it to several samples. A good estimator is: 1) one that produces a small difference between the mean of the estimates calculated from the samples and the parameter, known as the *bias*; and 2) one that has a small variance of sample estimates. These may be combined into a mean square error as the sum of the variance and square of the bias. Other important properties of a good estimators are: 3) *consistency*, in which both bias and variance tend to zero as n grows large; and 4) *robustness*, which is insensitiveness to departure from underlying assumptions. A characteristic used in evaluating a statistic is the number of *degrees of freedom* that the variable has. The mean calculated from from n observations is considered to have n degrees of freedom whereas the variance assumes that the mean is held constant so it has $n - 1$ degrees of freedom.

The field of statistics makes considerable use of theoretical probability density curves, $p(x)$, to represent its variables so that it can draw conclusions about the data with which it deals. If our data are assumed to be drawn from a continuous random variable then we may make the following statements about central tendency and variance. The median, m, is defined by

$$\int_{-\infty}^{m} p(x)\,dx = \int_{m}^{\infty} p(x)\,dx = 0.5, \tag{3.3}$$

and the mode is given by

$$\frac{dp(x)}{dx} = 0, \qquad \frac{d^2 p(x)}{dx^2} < 0. \tag{3.4}$$

Similarly the different parameters may then be written in terms of $p(x)$.

The mean, also known as the first moment, is the expected value of x, $E(x)$, given by

$$E(x) = \int_{-\infty}^{\infty} x p(x)\,dx = \mu_1. \tag{3.5}$$

The second moment, the variance, is the expected value of $(x - \mu_1)^2$,

$$E(x - \mu_1)^2 = \int_{-\infty}^{\infty} (x - \mu_1)^2\, p(x)\,dx = \mu_2. \tag{3.6}$$

The study of probability distributions was an outgrowth of interest in games of chance which can be traced back thousands of years but the mathematical theory of probability did not develop until the seventeenth century with the work of Huygens followed by many mathematicians such as Daniel Bernoulli and de Moivre (Kendall, 1956). Much of the early work centered on the chances of success of obtaining specific numbers in throwing dice. The result is the binomial distribution. In the limit, as the number of independent attempts increases, the binomial approaches one variously called the

Gaussian, Laplace, or *normal* distribution density function, $p_N(x)$, which is completely defined by the mean and the variance.

$$p_N(x) = \frac{1}{\sqrt{(2\pi\mu_2)}} \exp\left(-\frac{(x - \mu_1)^2}{2\mu_2}\right). \tag{3.7}$$

Examples of the normal curve are displayed in figures 3.6 and 3.10.

Other theoretical distributions also may be shown to approach the Gaussian in the limit. More importantly the *Central Limit Theorem* states that sample means are distributed normally regardless of the distributions within the individual samples. Also, a large number of observed variables display this form. Consequently, the normal curve is one that is frequently adopted for representing real data. This being the case and, in addition, if we can assume that the observations are *independent*, the data may be completely described by the population mean and variance. A large set of observations may thereby be reduced to two. In other words, from Equation (3.7), or from tables produced by that equation, we can calculate precisely the probability of the magnitude of the next observation, or the probability of exceeding a given magnitude.

Tables of the *cumulative normal distribution*, which give the area under the normal curve from $-\infty$ to positive values of x, which represents $\sqrt{\mu_2}$ or σ, for the standard normal distribution, are found in many statistical texts, e.g. (Kendall and Stuart, 1969; Mood and Graybill, 1963). Since the total area is 1.0 the area to the right of x is 1.0 minus the area to the left. Thus, in order to find the area between $x = -a$ and $x = +a$, we take the area at $+a$ and subtract the area to the left of $-a$. So, for the area between -1.0 and $+1.0$, we obtain $0.8413 - (1.0 - 0.8413) = 0.6826$. That says that, for the Gaussian distribution, 68.26 percent of all observations will be within $\pm\mu_2$ of the mean, or in other words, there is a 0.68 probability that any observation will be within $\pm\sqrt{\mu_2}$ of the mean. Similarly we may calculate that 95 percent of all observations will be within $\pm1.96\sqrt{\mu_2}$ of the mean, etc. Obviously the population parameters are unknown but depending upon how the frequency distribution is used, adjustments for the sample statistics may be made.

The concept of "independence" implies that the magnitude of any one observation is not influenced by the magnitude of another. For example, even though one person guesses that the temperature of a room is much higher than it actually is, the probability that the next person will guess similarly will be small, as given by the normal distribution. In fact we should expect that the next guess will be within $\pm1.96\sqrt{\mu_2}$ of the mean of all guesses 95 percent of time.

Unfortunately many data sets are non-normally distributed. Then, one solution is to transform the data into a normal distribution through the use of some mathematical relationship. A more direct approach is to calculate additional descriptors, such as higher moments of the distribution. One estimator of the third moment, $\hat{\mu}_3$, the average sum of the cubed deviations from the mean, is

$$\hat{\mu}_3 = \frac{1}{n} \sum_{i=0}^{n-1} (x[i] - \hat{\mu}_1)^3. \tag{3.8}$$

Similarly, higher moments may be calculated.

It should be mentioned here that different descriptive constants other than moments exist. In particular, *cumulants*, first introduced in 1937 by Cornish and Fisher (1937), are sometimes more useful from a theoretical viewpoint. Kendall and Stuart (1969) give conversion formulae between the first ten moments and cumulants. For a series having zero mean the first three are the same.

Now, if we return to our sample we see immediately that our distributions are not Gaussian. The precipitation data have an entirely different form and even our temperature data appear non normal (see the theoretical normal curve, superimposed in figure 3.6).

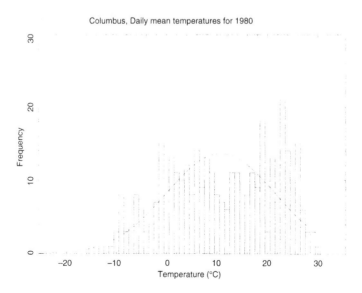

Figure 3.6 Normal curve superimposed on the histogram of figure 3.5

3.3 Two Variables

Before we go further with the single variable we need to jump ahead to consider the simple case of dealing with two variables. The reason will become apparent when we return to consider temporal and spatial data. The variables in this case are assumed to be in some way related. As an example let us consider the mean monthly temperatures for one location for two consecutive years, represented by x and y.

3.3.1 Covariance

As we have seen, a measure of the variability of one variable is the average of the squares of deviation from the mean. Extending that concept, we may estimate the covariability of two variables by calculating the average of the product of the deviations of the two variables from their respective means:

$$\hat{\rho}_2 = \frac{1}{(n-1)} \sum_{i=0}^{n-1} (x[i] - \hat{\mu}_{1,x})(y[i] - \hat{\mu}_{1,y}) \tag{3.9}$$

$$= \frac{1}{(n-1)} \left[\sum_{i=0}^{n-1} (x[i]\,y[i]) - n \frac{\left(\sum_{i=0}^{n-1} x[i]\right)}{n} \frac{\left(\sum_{i=0}^{n-1} y[i]\right)}{n} \right]. \tag{3.10}$$

If y be replaced by x it may be seen that these reduce to Equation (3.2). The $\hat{\rho}_2$ is an estimate of what is called the *covariance* of the two data sets. As will be shown in section 3.3.3, the covariance normalized (divided) by the product of the two variances gives us the square of the familiar *correlation coefficient*.

3.3.2 Regression

Another way of searching for a pattern of relationship is to plot the two variables one to one based upon some criterion: in this case, month. The result is what is known as a *scatter* diagram (see figure 3.7).

From looking at the distribution we might hypothesize some mathematically functional relationship between the two variables. The simplest is a linear relationship and the one applied here but other alternatives, many of which might be more appropriate, are available. Once the form of the function, in this case the straight line,

$$y[i] = u + v\,x[i] + \epsilon[i], \tag{3.11}$$

has been selected, how is the function or curve fitted: how are the u and v coefficients obtained from the observations? Again there are many alternatives

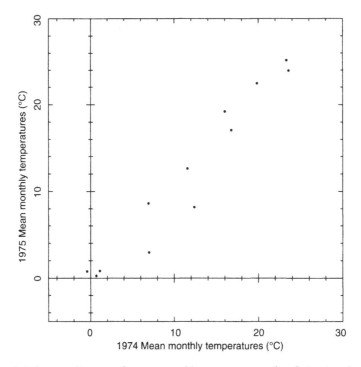

Figure 3.7 Scatter diagram of mean monthly temperatures for Columbus for 1974 and 1975. The data are paired based upon the month

but the method, originally introduced by Gauss, is now the standard procedure. This is to use the criterion of least squares. That is, the line defined by

$$\hat{y}[i] = u + v\,x[i], \tag{3.12}$$

is located so that the sums of squares of the differences between the observations and the line, ($\sum \epsilon^2[i]$), is a minimum. In other words, we seek the minimum of $\sum(y[i] - \hat{y}[i])^2$, where $y[i]$ refers to the observation at $x[i]$ and $\hat{y}[i]$ refers to the line at $x[i]$. The minimum may be found from setting its differential with respect to u zero and then with respect to v to zero (and testing that it is a minimum point). Explicitly, substituting from Equation (3.12) we have

$$(y[i] - \hat{y}[i])^2 = (y[i] - u - v\,x[i])^2. \tag{3.13}$$

Sum

$$\sum_{i=0}^{n-1} (y[i] - \hat{y}[i])^2 = \sum_{i=0}^{n-1} (y^2[i] - u^2 - v^2 x^2[i] - 2u\,y[i] - 2v\,y[i]\,x[i]$$
$$+ 2u\,v\,x[i])$$

$$= \sum_{i=0}^{n-1} (y^2[i]) + n\,u^2 + v^2 \sum_{i=0}^{n-1} (x^2[i]) - 2u \sum_{i=0}^{n-1} (y[i])$$

$$- 2v \sum_{i=0}^{n-1} (y[i]\,x[i]) + 2u\,v \sum_{i=0}^{n-1} (x[i]), \tag{3.14}$$

differentiate with respect to u and v, and we obtain

$$\frac{\partial \left[\sum (y[i] - \hat{y}[i])^2 \right]}{\partial u} = 2\,n\,v - 2 \sum (y[i]) + 2v \sum (x[i]), \tag{3.15}$$

$$\frac{\partial \left[\sum (y[i] - \hat{y}[i])^2 \right]}{\partial v} = 2v \sum (x^2[i]) - 2 \sum (y[i]\,x[i]) + 2u \sum (x[i]). \tag{3.16}$$

Now set each equal to zero for the minimum, rearrange and we obtain

$$n\,u + \sum_{i=0}^{n-1} (x[i])\,v = \sum_{i=0}^{n-1} (y[i]), \tag{3.17}$$

$$\sum_{i=0}^{n-1} (x[i])\,u + \sum_{i=0}^{n-1} (x^2[i])\,v = \sum_{i=0}^{n-1} (x[i]\,y[i]). \tag{3.18}$$

Equations (3.17) and (3.18) are known as the *normal* equations. Their solution is simple,

$$v = \frac{\sum_{i=0}^{n-1} (x[i]\,y[i]) - n\,\hat{\mu}_{1,x}\,\hat{\mu}_{1,y}}{\sum_{i=0}^{n-1} (x^2[i]) - n\,\hat{\mu}_{1,x}}, \tag{3.19}$$

$$u = \hat{\mu}_{1,y} - v\,\hat{\mu}_{1,x}. \tag{3.20}$$

Underlying the fitting of Equation (3.12) are the assumptions that $\epsilon[i]$ are normally distributed with mean zero and are uncorrelated and independent. These permit the estimation of confidence levels for the various calculated terms.

The normal equations and their solution may be generalized very nicely with matrix algebra (Draper and Smith, 1966). The general term applied to the process of relating y to x, even when x is composed of many variables,

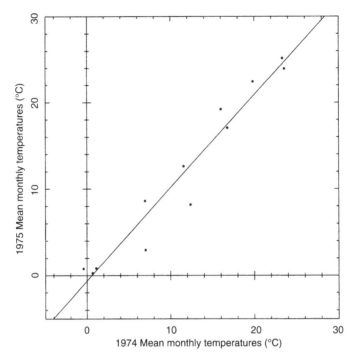

Figure 3.8 Regression line $\hat{y} = -0.6 + 1.1x$ obtained from plotted data

is known as *regression*. Originally the term was used by Galton to indicate certain relationships in the theory of heredity but its meaning has now broadened (Kendall and Buckland, 1971). The x variables are called "independent variables," "predicated variables," "predictors," or "regressors." y is called the "dependent variate," "predictand," or "regressand."

The process may be generalized for several independent variables. Then it is called *multiple regression*. There are several techniques for ranking the relative importance of the independent variables in affecting the dependent variable. Also, there are many computer statistical packages available that do all the processing. On the other hand, the user should not apply such techniques or packages blindly. Like any such technique a full understanding of the assumptions, the procedures involved, and the interpretation of the output should precede their use.

Using simple linear regression applied to the data in figure 3.7 we obtain $u = -0.6$ and $v = 1.1$ with the line plotted in figure 3.8. The u magnitude is the magnitude of y at $x = 0.0$ or where the line crosses the y axis. The v magnitude is the slope of the line, the tangent of the angle between the abscissa and the regression line. So far it has been assumed that y was being estimated from x, and we may explicitly indicate that by subscripting u and v as $u_{y \cdot x}$ and $v_{y \cdot x}$. We could also have assumed at the beginning that y were the independent variable and x were the dependent variable. In which case, for the data in figure 3.8, we would have $u_{x \cdot y} = 1.2$ and $v_{x \cdot y} = 0.88$. These define a different line which crosses the first at the means of the two variables $(\hat{\mu}_{1,x}, \hat{\mu}_{1,y})$.

3.3.3 Relationship Variables

The *covariance*, ρ_{xy}, is a measure of the covariability of x and y but it is in the units of the two variables. If the covariance is divided by the standard deviations of each variable then it turns out that we have a measure of their relationship on a scale of -1.0 to $+1.0$. This is usually represented by the letter r and is called the *correlation coefficient*. Its square, r^2, known as the *coefficient of determination* or just "r square," may be thought of as measuring the percentage variance of y that is contributed by x, or vice versa,

$$r^2 = \frac{\rho_{xy}^2}{\mu_{2,x} \, \mu_{2,y}}. \tag{3.21}$$

A review of Equation (3.19) reveals that $v_{y \cdot x}$ is the covariance divided by the variance of x,

$$v_{y \cdot x} = \frac{\rho_{xy}}{\mu_{2.x}} \tag{3.22}$$

Similarly

$$v_{x \cdot y} = \frac{\rho_{xy}}{\mu_{2.y}}. \tag{3.23}$$

Therefore

$$r^2 = v_{y \cdot x} \, v_{x \cdot y}. \tag{3.24}$$

When the regression lines coincide $r = 1.0$.

By their nature the various statistics, that are being calculated to show a connection between variables, are only estimates of the population values. Therefore, many procedures exist for assessing whether the resulting numbers are reasonable estimates for the parameters that they represent. This is a large and important area but it will not be discussed here. It should also be noted that regardless of how confident we might be in our statistical estimates they are only generated by manipulating the number pairs. They may be suggestive of physical connections but do not prove that they exist. The "suggestions" should be subsequently investigated by studying the physical processes.

3.4 Dependence

3.4.1 Auto-covariance

If we return now to reconsider atmospheric time series we realize that the assumption about independence cannot be accepted. There is a memory in the sequence. The temperature at one time is likely to be similar to those before and after it. Depending upon the time interval this similarity will vary. From figure 3.3 we can see that if the daily mean temperature is low on one day, say $-5°C$, the chances are high that the mean temperature will also be about $-5°C$ the next day. That is contrary to what we expect with independent data where the temperature the next day should be within approximately two standard deviations of the mean 95 percent of the time, that is for these data between about 0 and 22°C. In addition, because of the annual oscillation of incoming solar radiation, we expect a close similarity of temperatures one year apart. As a consequence, in order to summarize our data more fully, we do need to document this interdependence. One way to do that is to calculate the lagged auto-covariances. That is, instead of summing the product of the deviation of the variable from the mean with itself, we sum the product of the deviation of the variable (mean removed) with itself at lagged intervals, say p apart. Such a series, represented by ρ_{xx}, will be a function of p. One estimator of the lagged covariance is

$$\hat{\rho}_{xx}[p] = \frac{1}{n - |p|} \sum_{i=0}^{n-1-|p|} (x[i] - \hat{\mu}_1)(x[i + p] - \hat{\mu}_1). \qquad (3.25)$$

To make the equations simpler we shall assume that all series from this point on have zero means. Then Equation (3.25) becomes

$$xx[p] = \hat{\rho}_{xx}[p] = \frac{1}{n - |p|} \sum_{i=0}^{n-1-|p|} x[i]x[i + p]. \qquad (3.26)$$

Notice that p may be negative so that this is a symmetrical function running from $-(n-|p|)$ to $(n-|p|)$. Sometimes $\hat{\rho}_{xx}[p]$ is normalized with the variance $(\hat{\mu}_2)$, which is also the magnitude of $\hat{\rho}_{xx}[0]$,

$$\hat{\gamma}_{xx}[p] = \frac{1}{\hat{\mu}_2} \frac{1}{(n - |p|)} \sum_{i=0}^{n-1-|p|} x[i]x[i + p]. \qquad (3.27)$$

This function, known as the *auto-correlation* function, starts with a magnitude of 1.0 at $p = 0$ and descends to oscillate around zero. Unfortunately, just as the raw data that we are trying to characterize are not independent, the auto-covariances are not either. What we need is some mathematical transformation that will give us an independent series. There are several such functions to chose from but one also turns out to have a valuable physical interpretation. This is the Fourier series.

Before we review that topic we should recognize that single temporal and spatial series are not ensembles. Therefore, if we are to use many of the techniques of statistics that apply to ensembles, we have to make assumptions about our series. Typically we have only one series: one realization of the variable for a point in time or space. This might be one observer taking cloud observations four times a day. Let us use upper case letters to identify members of the ensemble and lower case letters to identify observations in time or space. Then $x[j, J]$ will represent an observer $[J]$ at point $[j]$ in the series. There will be N people each observing n times. Now we may calculate our statistics in two ways. First, we may calculate them by letting $[J]$ vary (keeping $[j]$ fixed), giving us the ensemble statistics, and then consider what happens to them as $[j]$ changes. Secondly, we may calculate allowing $[j]$ to vary and then see how they change as $[J]$ changes.

If the assumption may be made that the ensemble mean and auto-covariances do not change significantly as $[j]$ changes then the series is said to be *weakly stationary*. If higher moments also do not change significantly then the ensemble is *strongly stationary*.

If the series means and auto-covariances (calculated from one realization) do not change significantly as $[J]$ changes then the ensemble is said to be *ergodic*. Ergodicity presupposes stationarity.

There is some debate in the literature over how important it is to have ergodic data. Clearly as we deal with atmospheric phenomena undergoing climatic change these assumptions will be stretched. As in many practical situations the required statistical assumptions may not apply. In such cases the calculated statistics may lead to misleading conclusions.

3.4.2 Fourier Series

The basis of oscillating trigonometric functions has a long history. Pythagoras (ca. 530 BC) had recognized the importance of harmonics in music and Kepler at the end of the sixteenth century extended the idea of harmony to the motion of the planets. The first mathematical treatment of wave motion was given by Newton in 1687 and was included in *Principia* (Newton, 1687). The equations, now known as Fourier series were first used by Daniel Bernoulli who worked on vibrating strings as early as 1728. Lagrange also discussed the technique in 1772 and, as result, was negative about subsequent work (see section 2.8.2). The theoretical basis of the idea that almost any variable could be represented as the sum of sines and cosines was first announced by Jean Baptiste Joseph Fourier in 1807 and published in full in 1822 (Fourier, 1822). Subsequently

Gustav Dirichlet (1805–59), Henri Lebesgue (1875–1941) and Lipót Fejér (1880–1959) developed rigorous proofs of Fourier's work. Essentially Fourier showed that a sequence of n data could be replaced by another sequence of n numbers that specified a set of sines/cosines of increasing frequency. In turn these n numbers could return exactly the original n data. The mathematical requirements are that the data $x[t]$ be single valued (i.e. there can only be one mean temperature for a specified place on a specified day), that $x[t]$ be defined for every point between the beginning and the end, and that there be a finite number of maxima, minima, and discontinuities.

Figure 3.9 displays a cosine curve (dash-dot) and a sine curve (dot). The equations for these are

$$y_c = a[k]\cos(kx) \tag{3.28}$$
$$y_s = b[k]\sin(kx) \tag{3.29}$$

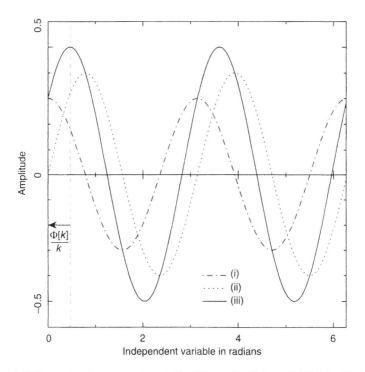

Figure 3.9 Curves for frequency $k = 2$. (i) $a[k]\cos(2\pi jk/n)$, (ii) $b[k]\sin(2\pi jk/n)$, and (iii) $A[k]\cos\{(2\pi jk/n) - \Phi[k]\}$. Curve (iii) is the sum of curves (i) and (ii). It is a cosine curve which begins at $-\Phi[k]/k$ radians from the crest

The independent variable, x, in radians varies from 0 to 2π, which we call the total length. If we have n discrete observations spaced Δt apart, then $n\Delta t = 2\pi$. The jth observation will be $(j - 1)\Delta t$ from the origin. The variable k indicates how many times the curve completes a cycle in the total length. It is called the *frequency*. Sometimes it is useful to view frequency in terms of *wavelength*, the distance between similar consecutive positions in our wave, e.g peak to peak. Wavelength, λ, corresponding to frequency k is $(2\pi)/k$ radians or $(n\,\Delta t)/k$ data units. In figure 3.9 $k = 2$. The variables a and b are the distances of the peaks of the respective curves from the abscissa. They are called *amplitudes* and are labeled by the frequency $[k]$ to which they belong. In our example $a = 3$ and $b = 4$.

The cosine peaks at 0 whereas sine peaks at $\pi/2$. If the cosine and sine are added together we have a third sinusoid peaking at some other location specified by the ratio of a and b. To generalize, a sinusoidal curve at a specific frequency k may be represented, either as the sum of a cosine and a sine each with their own amplitudes, $a[k]$ and $b[k]$ respectively, or as just a sinusoid, say a cosine, with an amplitude $A[k]$ and a *phase angle* $\Phi[k]$,

$$y[j] = a[k]\cos\left(\frac{2\pi jk}{n}\right) + b[k]\sin\left(\frac{2\pi jk}{n}\right) \tag{3.30}$$

$$= A[k]\cos\left\{\left(\frac{2\pi jk}{n}\right) - \Phi[k]\right\}, \tag{3.31}$$

where j is the data-domain counter,

$$A^2[k] = a^2[k] + b^2[k], \tag{3.32}$$

and

$$\tan \Phi[k] = \frac{b[k]}{a[k]}. \tag{3.33}$$

Substitution of a and b from our example into Equations (3.32) and (3.33) gives $A[2] = 5$ and $\Phi[2] = 0.927$. The phase angle, when divided by the frequency, $\Phi[k]/k$, specifies the distance from the origin that the nearest crest of the wave occurs. This is called the *phase shift*. To convert to data-domain units $\Phi[k]/k$ must be multiplied by $n\,\Delta t/2\pi$. The curve for Equation (3.31) is plotted as a continuous line in figure 3.9. The phase shift is 0.464.

The *Fourier Series* is defined as the sum of the terms in either Equation (3.30) or Equation (3.31) over all frequencies k.

$$y[j] = \sum_{k=0}^{n/2} \left[a[k] \cos\left(\frac{2\pi j k}{n}\right) + b[k] \sin\left(\frac{2\pi j k}{n}\right) \right] \qquad (3.34)$$

$$= \sum_{k=0}^{n/2} \left[A[k] \cos\left\{\left(\frac{2\pi j k}{n}\right) - \Phi[k]\right\} \right], \qquad (3.35)$$

k may take on the integer magnitudes of zero to $n/2$ for n even or $(n-1)/2$ for n odd. In Equation (3.34) $a[0]$ is the mean of the series and there is no $b[0]$. Also, if n is odd, there is only an $a[(n-1)/2]$ no equivalent b. In fact, n observations will produce n coefficients, as plus bs or As plus Φs. The method of estimating the coefficients is least squares but because trigonometric functions are orthogonal the arithmetic turns out to be very simple. In matrix terms, the off-diagonal elements of the normal equations are zero. Yet another feature is the recurrence of the same magnitudes over and over again within the series. That leads to a further simplification of calculations known as the *Fast Fourier Transform algorithm* (Gentleman and Sande, 1966; Rayner, 1973b). Also, unlike the fitting of a straight line as performed in section 3.3.2, the fitting of Fourier Series is exact. The sum of the sinusoids passes through every point. As a result, no information is lost when a series is converted to cosine and sine coefficients. Therefore, the original series may be reconstructed from the coefficients by Equation (3.34) or Equation (3.35). The distribution of elements (coefficients, amplitudes, phases, etc.) according to frequency is called a *spectrum*.

The equations for obtaining $a[k]$ and $b[k]$ are

$$a[k] = \frac{2}{n} \sum_{j=0}^{n-1} y[j] \cos\left(\frac{2\pi j k}{n}\right) \qquad (3.36)$$

and

$$b[k] = \frac{2}{n} \sum_{j=0}^{n-1} y[j] \sin\left(\frac{2\pi j k}{n}\right). \qquad (3.37)$$

When $k = 0$ and $k = n/2$ (for n even) the coefficients are $1/n$ not $2/n$. In order to avoid these exceptions we may calculate the ks from $-n/2$ to $+n/2$. Then all the coefficients are $1/n$. Also, for symmetry in future equations, our data are defined to run from $-n/2$ to $+n/2$ instead of from 0 to n. Then Equations (3.36) and (3.37) may be rewritten with a tilde over the amplitudes,

$$\tilde{a}[k] = \frac{1}{n} \sum_{j=-n/2}^{n/2} y[j] \cos\left(\frac{2\pi jk}{n}\right) \tag{3.38}$$

and

$$\tilde{b}[k] = \frac{1}{n} \sum_{j=-n/2}^{n/2} y[j] \sin\left(\frac{2\pi jk}{n}\right). \tag{3.39}$$

Since, when $k = 0$, all cosine terms are 1, $\tilde{a}[0]$ is equal to the mean of the series. The $\tilde{b}[0]$ term is always zero. In the following it is assumed that the series have zero mean.

It should be noted that the $\tilde{a}$s (cosine coefficients) are symmetrical around the origin $k = 0$ so $\tilde{a}[k] = \tilde{a}[-k]$, whereas the $\tilde{b}$s (sine coefficients) are anti-symmetrical, $\tilde{b}[k] = -\tilde{b}[-k]$. Sometimes the $\tilde{a}$s are said to be even and the $\tilde{b}$s odd.

A useful by-product of representing data with Fourier coefficients is that they subdivide the variance. The variance of one sinusoid at frequency k is given by

$$\text{var}[k] = A^2[k]/2 = 2(\tilde{a}^2[k] + \tilde{b}^2[k]) \tag{3.40}$$

and the sum of all the variances is equal to the variance of the $y[j]$ series. In other words, Fourier series break down the total variance into different frequency or period components. This breakdown is therefore the *variance spectrum*.

An important assumption not yet mentioned is that $y[j]$ is periodic. That is, $y[j]$ repeats itself both forwards to $+\infty$ and backwards to $-\infty$ from the "total length." In many situations, for example where simple curve fitting is required, this may not be a problem. If, however, the coefficients are to be used in some form of inference, that is that they represent the variance at different frequencies, then the validity of the assumption is critical. Examples of periodic functions are earth motion and observations made instantaneously around a latitude circle. As an illustration, Peixoto et al. (1964) have published the breakdown of topography along parallels of the earth. It should be noted too that with many series most of the variance is concentrated in the lower frequencies so that for such data they may be adequately represented by a few components, far fewer numbers than in the original series (Moellering and Rayner, 1981).

Diurnal and annual fluctuations of natural phenomena often approach periodicity. In contrast, other weather fluctuations are not periodic. A solution in such situations is outlined in section 3.4.5.

3.4.3 Complex Representation

Since $\tilde{a}$ and $\tilde{b}$ are a number pair they may be represented by a complex number as introduced in section 2.7.3. Because these Fourier coefficients are derived from the $y[j]$ series we shall use $Y[k]$ to represent the complex combination of them,

$$Y[k] = \tilde{a}[k] - i\tilde{b}[k].\tag{3.41}$$

The fitting could have been performed in the reverse direction in which case the $\tilde{a}$s would have remained the same, whereas the $\tilde{b}$s would have changed sign and we obtain the complex conjugate,

$$Y[-k] = \tilde{a}[-k] + i\tilde{b}[-k] = Y^*[k]\tag{3.42}$$

Notice the convention used here for the signs within Y. Also, from Equations (2.93) and (2.94) cosines and sines may be represented by complex numbers so our Fourier Series may now be written in general form

$$Y[k] = 1/n \sum_{j=-n/2}^{+n/2} y[j]e^{-i2\pi jk/n},\tag{3.43}$$

and

$$y[j] = \sum_{k=-n/2}^{+n/2} Y[k]e^{i2\pi jk/n}.\tag{3.44}$$

3.4.4 Fourier Transform and Convolution

One approach to the analysis of a sample section of data taken out of a continuum is to assume that the sections before (to $-\infty$) and after (to $+\infty$) are zero (with the sample having zero mean). Since n goes to ∞, the coefficient in Equation (3.43) (i.e. $1/n$) becomes zero. To avoid this problem Equation (3.43) is multiplied by n which is absorbed into the $\tilde{a}[k]$s, $\tilde{b}[k]$s, and $Y[k]$s which are relabeled $\tilde{a}[f]$s, $\tilde{b}[f]$s, and $Y[f]$s. They represent density functions: analogous to probability densities or intensities given by the Planck radiation equation. The Fourier Series Equations (3.43) and (3.44) for integer frequencies at k and equally spaced observations at j now become continuous functions in f and t, and integration is used instead of summation over the range $-\infty$ to $+\infty$,

$$Y[f] = \int\limits_{t=-\infty}^{+\infty} y[t]e^{(-i2\pi tf)}\,dt, \tag{3.45}$$

$$y[t] = \int\limits_{f=-\infty}^{+\infty} Y[f]e^{(i2\pi tf)}\,df. \tag{3.46}$$

These equations describe *Fourier transforms*. If we review the process of calculation we see that we may continue to use Equations (3.38) and (3.39) for the Fourier transform but we must consider the estimate, say variance, not coming from an integer frequency k as in the periodic case but as a total coming from an elementary band of width $1/n\,\Delta t$ centered on k.

If $y[f]$ (or $y[j]$) is multiplied by another function, say $h[f]$ (or $h[k]$) term by term, the Fourier transform of the product is given by

$$\int\limits_{-\infty}^{+\infty} y[t]h[t]e^{(-i2\pi tf)}\,dt = \int\limits_{-\infty}^{+\infty} Y[f]H[f_1 - f]\,df = X[f] * H[f], \tag{3.47}$$

where the operation on $X[f]$ and $H[f]$, denoted by an asterisk, $*$, like a multiplication sign, is called a *convolution*.

The comparable operation in the data-domain is typically called *filtering*. For example, one form of filtering is the application of a weighting function, $w[t]$, whose integral over all t is unity. It is

$$x_1[t] = \lim_{T \to \infty} \frac{1}{T} \int\limits_{-\infty}^{+\infty} x[t_1 + t]w[t]\,dt. \tag{3.48}$$

The transform of x_1 is the product of the transform of x and the conjugate of the transform of $w[t]$,

$$\int\limits_{-\infty}^{+\infty} \left(\int\limits_{-\infty}^{+\infty} x[t_1 + t]w[t]\,dt \right) e^{(-i\pi ft_1)}\,dt_1 = X[f]W^*[f]. \tag{3.49}$$

As before, the $*$ as a superscript, denotes conjugate.

Note that, if $w[t]$ be replaced by $x[t]$ and we average, we have the auto-covariance function on the left hand side (Equation 3.26), and the right hand side is the sum of the squares of the cosine and sine coefficients: the spectrum of the variance,

$$\int_{-\infty}^{+\infty} \left(\lim_{T \to \infty} \frac{1}{T} \int_{-\infty}^{+\infty} x[t_1 + t] x[t] \, dt \right) e^{(-i\pi f t_1)} \, dt_1 = \int_{-\infty}^{+\infty} xx[t_1] e^{(-i\pi f t_1)} \, dt_1$$

$$\int_{-\infty}^{+\infty} xx[t] e^{(-i\pi f t)} \, dt = \lim_{T \to \infty} \frac{1}{T} (X[f] X^*[f]) \tag{3.50}$$

$$= \lim_{T \to \infty} \frac{1}{T} |X[f]|^2, \tag{3.51}$$

which is proportional to the transform of the raw data squared.

3.4.5 Spectral Analysis of Non-periodic Functions

Excellent texts exist on this topic so only a brief outline follows. It cannot do justice to this field for, as stated by Kay (1988), the author of the most widely used book in the 1990s, "it is an impossible task to describe the vast field of modern spectral estimation in a single book."

Kay's book, which uses 545 pages plus computer disks, looks at the various estimators for the spectrum with calculated examples. Here we shall concentrate upon, what Kay calls, the classical approach. It is perhaps the easiest to understand and to calculate. Early references to the classical method include Blackman and Tukey (1958), which was followed later by others (Jenkins and Watts, 1968; Brillinger, 1975). A simple exposition with exercises and examples from geography and climatology is by Rayner (1971).

Now, as already stated, sines and cosines are orthogonal. In other words, each element in the series is independent of another. That fulfills the requirement of independence, something we were looking for in section 3.4.1. However, Fourier Series are periodic which seriously limits their simple use on many natural series which are clearly non-periodic.

Preparation of the Data

Even though atmospheric time series in general are non-periodic, some components, such as the diurnal and annual fluctuations are related to planetary motion which, as Kepler discovered, can be described by harmonics. For our temperature data in figure 3.3, while the diurnal variation is clearly visible, it has been removed by averaging the maxima and minima, but the annual oscillation remains. If this were the only component of variation present, the lagged auto-covariances would give us large negative peaks at

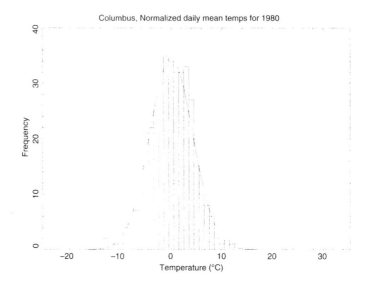

Figure 3.10 Frequency plot of the temperature data in figure 3.3 with the mean and annual oscillation removed

1/2, 3/2, 5/2, ..., and large positive peaks at 1, 2, 3, ... years. Because these will blur other frequency components nearby, it is a good idea to remove them early in the analysis. This can be done by fitting a cosinusoid (Equation 3.31) to the raw series. In the example given for Columbus in 1980 the amplitude turns out to be $13.94°C$ accounting for a variance of $97.27°C^2$ (Equation 3.40) which is 84.4 percent of the original total, and the phase -161 days (July 25). Once the annual variation has been removed we see that the remaining distribution is much more nearly normal (figure 3.10).

Preliminary analysis will usually reveal any such important trends that exist so it is logical to deal with those separately. Furthermore, as has just been illustrated, they should be removed since they tend to confuse and mask the underlying series.

Windows

As we saw in section 3.4.4 one way of handling non-periodicity is to allow the the series outside the limit of observation to take on the magnitude of zero (i.e. the mean of the observed sample). Such an assumption would appear to be no better than periodicity until we consider looking at our sample as being seen through a window. Analogous to the ordinary window, a data-domain window allows us to see only a small piece of space beyond. The walls around the window are equivalent to multiplying what is beyond it by zero. In Fourier terms this is described by Equation (3.47) where $h[t]$ is the window, being equal to $+1$ when it coincides with our observations and zero elsewhere. That means our spectrum is a convolution of the data with the window.

Unfortunately a regular window has some nasty properties when it comes to frequency analysis. A rectangle, for that is what a simple "do nothing" $h[k]$ is, requires a very large number of cosinusoidal curves to describe it. In the convolution it combines significant amounts of the variable from selected higher frequencies with the lower frequencies. Therefore it is not a good choice. Other windows that gradually change from $+1$ to 0 are far better for spectral work. Reference may be made two small books by Barber (1961) and by Jennison (1961) on transforms that provide a large number of examples of the relationship between simple data-domain shapes and their spectral counterparts. In the classical approach the "Hanning" window, $H^*[f] = 0.25, 0.50, 0.25$, named after the German meteorologist, Julius von Hann, is often used.

Practical Calculation

Having removed the mean and trends then, we may subject our reduced data to auto-covariance calculations, a Fourier transformation, and a convolution with a suitable window function. The auto-covariance function can only be calculated for fewer and fewer terms as the lag increases. Therefore we limit the covariance function to about $m = n/10$. Because the auto-covariance function is symmetrical about $k = 0$ we need only calculate $+k$ values to $m/2$ and then use only Equation (3.38). The result is the variance spectrum of our sample, $y[t]$.

Alternatively it is unnecessary to calculate the auto-covariance function. As revealed by Equation (3.51) we may calculate the spectrum directly from the series that has had the mean and annual cycle removed. The window is now applied in the data-domain. An obvious choice is the cosine bell applied at the beginning and at the end (Tukey, 1967). This is called *tapering*. Then the individual frequency bands are summed over some selected interval to provide reasonable degrees of freedom on the estimates.

Applied to the reduced data of figure 3.3 we obtain the variance spectrum displayed in figure 3.11. This diagram shows that there is little variance in the frequency band centered on 0, the one that includes fluctuations of periods greater than 160 days. Table 3.1 presents a numerical overall breakdown. Bands 1 to 3 (23–160 days) account for 31.6 percent of the variance in this spectrum or 4.9 percent of the total. Bands 4 to 7 (11–23 days) and 10 to 14 (5.5–8.5 days) account for another 21.2 percent (3.3 percent of total) and 23.2 percent (3.6 percent of the total) respectively. These may be compared with the 84.4 percent of the total in the annual period. The diurnal oscillation would also have accounted for a

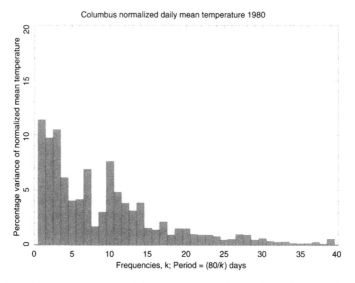

Figure 3.11 Variance spectrum of the temperature data in figure 3.3 with the mean, diurnal, and annual components removed

Table 3.1 Statistics from the Columbus temperature data

	Magnitude	%	%
Mean	$10.7^\circ C$		
Variance			
Total	$115.24^\circ C^2$	100.0	
Annual Cycle	$97.27^\circ C^2$	84.4	
Annual Removed	$17.97^\circ C^2$	15.6	100.0
23–160 days		4.9	31.6
11–13 days		3.3	21.2
5–8.5 days		3.6	23.2
Remaining		3.8	24.0

large percentage but it has been included neither in the total nor in the percentages.

As with other statistical estimates we need to give some measure of confidence that we have in our estimates. If all the assumptions are valid, variance spectral estimates are distributed as chi square with degrees of freedom given approximately by the number of observations less the number of observations smoothed at one end by the cosine bell all divided by the number of spectral estimates.

3.4.6 Non-normal and Dependent Data

Independent data that are non-normally distributed require higher moments for their description. In addition, dependent data require higher order spectra.

For example, an estimate of the third moment M^3 about the mean is the average of the sum of cubes of deviations of the observations from their mean. Assuming that $y[j]$ has zero mean and applying Equations (3.41) and (3.44) we obtain

$$nM^3 = \sum_{j=0}^{n-1} (y[j])^3 \qquad (3.52)$$

$$= nD \sum_{k_1=1}^{n/4} \sum_{k_2=k_1}^{n/2-k_1} Y[k_1]Y[k_2]Y^*[k_3], \qquad (3.53)$$

where $D = 12$ when $k_1 \neq k + 2$ and $D = 6$ when $k_1 = k_2$. The algebra involved is substantial but most terms drop out with summation. The only ones to remain are those whose frequencies sum to zero, i.e. $k_1 + k_2 + k_3 = 0$ or $-k_3 = (k_1 + k_2)$, where the conjugate frequency is equal to the sum of the two positive frequencies. When Equation (3.53) is expanded it contains

$$\left\{ \left(\hat{a}[k_1]\hat{a}[k_2] - \hat{b}[k_1]\hat{b}[k_2] \right) - i\left(\hat{a}[k_1]\hat{b}[k_2] + \hat{a}[k_2]\hat{b}[k_1] \right) \right\}$$
$$\times \left\{ \hat{a}[k_3] + i\hat{b}[k_3] \right\}.$$

The hats indicate that these as and bs are estimates that have been calculated using Equations (3.38) and (3.39) after tapering has been applied. If the data $y[j]$ were non-periodic a windowing would have been applied and the ks would have referred to the centers of elementary bands. The interaction of two waves of unlike frequencies ($y[k_1]$, $y[k_2]$) produces a third wave of frequency k_3, i.e. $y[k_3]$, given by the first curly brackets in above expression. The distribution of this wave, $y[k_3]$, is symmetrical but when another wave $y'[k_3]$ of the same frequency k_3, given by the second curly brackets, interacts with it the result is asymmetric: skewed. Each triple product $Y[k_1]Y[k_2]Y^*[k_3]$ in the double sum represents a separate contribution to the total third moment. Because k_1 and k_2 control k_3 it is sufficient to label elements in the spectrum with k_1 and k_2, for example $M^3[k_1, k_2]$, and to plot them on a two-dimensional diagram. Also, because of the symmetries involved only a triangular region need be plotted with k_1 allowed to vary from 1 to $n/4$ and k_2 from k_1 to $n/2 - k_1$. A large magnitude of M^3 will indicate that three different scales are interacting strongly.

The spectrum of the third moment is known as the *bispectrum*. Following a similar argument we may arrive at the spectra of higher order moments. They are collectively called *polyspectra* or kth order spectra (Brillinger and Rosenblatt, 1967). The description above is written in terms of periodic functions but it is a simple step to develop the equivalent equations for the non-periodic case.

3.4.7 Dependence Summary

Data from the atmosphere in general are not independent and they are often not normally distributed. Therefore, their statistical description requires more than just the mean and the variance. Dependence calls for the auto-covariance function. Because the auto-covariance function is itself not independent an orthogonal transformation is helpful in the interpretation of the results. The Fourier transformation as a choice has significant advantages because it represents a scale breakdown of atmospheric phenomena: it has a physical as well as a statistical interpretation. Extension of this approach for the description of non-normality leads us to use polyspectra.

3.5 Dependence for More Than One Variable

3.5.1 Two Entities

In section 3.3.2 we considered the linear relationship between two variables. Although not stated there, the general assumption was that the monthly temperatures were from independent ensembles. In fact, we might expect dependence in such data. That means that $\hat{y}[i]$ in Equation (3.12), $\hat{y}[i] = u + v\,x[i]$, is affected not just by x at i but also by x at $i-1, i-2, i-3$, etc. and possibly at $i+1, i+2, i+3$, etc. Incorporating such dependencies into the regression equation in a most general form we have

$$\hat{y}[t] = u[t] + \int\limits_{-\infty}^{\infty} v[q]x[t+q]\,dq. \qquad (3.54)$$

This equation may be solved as a function of frequency. First, both sides are multiplied by $x[t+p]$ and averaged over all t. If $x[t]$ and $u[t]$ are uncorrelated that term is zero and we are left with

$$\lim_{T\to\infty} \frac{1}{T} \int\limits_{-\infty}^{\infty} y[t]x[t+p]\,dt = \int\limits_{-\infty}^{\infty} v[q] \lim_{T\to\infty} \int\limits_{-\infty}^{\infty} x[t+q]x[t+p]\,dt\,dq,$$

$$(3.55)$$

or, if we introduce $xy[p]$,

$$xy[p] = \int_{-\infty}^{\infty} v[q]\,xx[p-q]\,dq. \tag{3.56}$$

Transform and rearrange so

$$V[f] = \frac{XY[f]}{XX[f]}. \tag{3.57}$$

This is the frequency response function of the system and is analogous to v in Equation (3.22),

$$v = \frac{(\sum xy)/n}{(\sum x^2)/n}.$$

Similarly the "r square" function of Equation (3.24) becomes

$$R_y^2 x[f] = \frac{XY^2[f]}{XX[f]\,YY[f]} \tag{3.58}$$

and is known as the *coherence* or *coherency*.

In many situations in meteorology the products of series are required. For example, the transport of quantities such as water vapor is directly related to the product of specific humidity and the wind vector. Both specific humidity and wind are variables measured in space and in time. The scale breakdown of such transport is given by $XY[k]$. Following a similar procedure as outlined in section 3.4.5 we may write down the actual equations for calculating this function. Instead of using the lagged covariances $xy[p]$ we may apply a window and transform each series separately using Equations (3.38) and (3.39) producing estimates (hatted) of the as and bs for each elementary band. Then we write them in complex form. The result is

$$XY[k] = \left(\frac{\hat{a}_x[k]\hat{a}_y[k] + \hat{b}_x[k]\hat{b}_y[k]}{2}\right) - i\left(\frac{\hat{a}_x[k]\hat{b}_y[k] - \hat{a}_y[k]\hat{b}_x[k]}{2}\right) \tag{3.59}$$

$$= XY_E[k] - iXY_O[k]. \tag{3.60}$$

Finally the elementary bands need to be summed over a wider interval as was done with the univariate spectral estimates.

The real part of (3.60), XY_E, is even or symmetrical about $k = 0$, and is known as the cospectrum. This is the actual scale breakdown of the product $x[j]y[j]$. The odd or asymmetrical part of (3.60) is known as the *quadrature spectrum* and is used with the cospectrum to estimate the average phase difference between the oscillations in x and y at each frequency band. The phase controls the effectiveness of the system. For example, if isobars and isotherms are exactly in phase around a latitude circle no net sensible heat will be transported by the geostrophic wind across that circle. Synoptic systems must display asymmetry to produce net transports.

3.5.2 The Transfer Function

Natural and artificial systems take in variables and put out variables. For example, a mercury thermometer absorbs thermal energy (input) and displays a change in the length of the column of mercury (output). Many people assume a one to one correspondence between the air temperature and the mercury length. This would be suggested by simple linear regression (Equation 3.12). In fact, the temperature of the air may vary quite rapidly, too fast for the mercury to respond. In order to estimate the way in which the instrument modifies the input it may be exposed to a controlled set of fluctuations which are compared to the output with Equation (3.57). The result is known as the *frequency transfer function*. Depending upon the use of the output a particular response may or may not be acceptable. A calibrated mercury thermometer is ideal for representing the general temperature over several minutes. It is a very poor instrument for estimating vertical sensible heat fluxes by atmospheric turbulence taking place within a few seconds because it does not register the higher frequencies.

Most systems are more complex than the thermometer but they may be analyzed in a similar way. Typically we use the outputs as the bases for our theories about nature but those outputs are always in some way filtered and may be misleading. Clearly, therefore, it is valuable to know what is being transferred: what the transfer function is, part of which is the frequency response function.

Sometimes we use filters intentionally. For example, many time series contain a lot of high frequency oscillations that tend to mask the longer term trends. Hence the data are often processed to remove this "noise." Unfortunately a large number of researchers continue to apply a running mean for this purpose despite an excellent discussion by Holloway explaining its serious drawbacks (Holloway, 1958). The transfer function shows that not only does the running mean not remove some high frequencies it even changes their phase by π. Thus if the diurnal cycle were one of these frequencies the expected nightly minima and daylight maxima would be interchanged.

3.5.3 Three Entities

As early as 1894 Osbourne Reynolds (see section 8.7) showed that the rate of change of kinetic energy in a fluid was a function of the third moment and of the triple products of the components of flow (Reynolds, 1894). This was extended as a scale analysis to the larger components of atmospheric flow by Saltzman (1957) and to oceanic flow by Hasselmann et al. (1963). The energy flux between the scales is given by Equation (3.53) where two of the Ys are the zonal wind and the other is the meridional component.

Clearly in order to understand the general circulation of the atmosphere, which is intimately interrelated to climate, we need to estimate, among other components, the energy transfers between the different scales of motion. Bispectral techniques provide means of doing that.

3.6 Comment

This chapter began with the argument that we need to summarize the vast quantities of data that are being generated continuously by observation and by modeling. Statistics gives us the tools to do that although traditionally we have employed very few of them. The mean, minimum and maximum are the most frequently used. Second, we need to calculate quantities relevant to understanding climate. For example, we might wish to estimate how much water moves from one region to another and by what systems. Again, statistics can provide the answer as a scale breakdown, which in turn may often be interpreted in terms of the physical system. Third, in our search for explanation, statistics can be used to identify possible connections or relationships that can subsequently be investigated by the physics of the situation. Finally, since we shall be using theoretical numerical models, we shall need to evaluate their performance relative to observations. Sophisticated statistical procedures are available for that important task. Like mathematics, statistics is a field basic for advanced work in climatology.

Surprisingly, despite the enormous amounts of data available and the recognition that more than the mean is required to describe climate very few statistics are published. Textbooks typically display only mean global maps for January and July. Observed extremes may also be given but maps of even the variance are seldom to be found.

Chapter 4

Mechanics

4.1 Newton's Definitions and Laws

By the seventeenth century considerable research had been conducted on forces and motions especially as they related to levers, projectiles, falling bodies, pendulums and planetary motions. The development of science proceeded despite strong opposition from the establishment, especially the Roman Catholic Church, where the new ideas contradicted existing dogma. The theory expounded by the Polish astronomer, Nicholas Copernicus (1473–1543), who had studied and lectured in Italy, that the sun was the center of our local universe rather than the earth, was of particular concern. The German, Johannes Kepler (1571–1630), who had become the assistant to the Danish astronomer Tycho Brahe (1546–1601) in 1600, further refined this heresy. He formulated his three laws of planetary motion: that (i) the orbits of the planets were ellipses with the sun in a common focus, (ii) the line joining a planet to the sun swept out equal areas in equal times, and (iii) the squares of the periodic times were proportional to the cubes of the mean distances from the sun.

It was not unexpected then that, when the Italian Galileo Galilei (1564–1642) published statements supporting a solar system theory, he was ordered in 1616 to abandon these views. Galileo spent most of his life in Pisa and Florence. He was a professor first at Pisa for two years and then at Padua for eighteen. In 1610 he returned to Florence as mathematician and philosopher to the grand duke of Tuscany and chief mathematician with no teaching duties at Pisa. Due to his continued adherence to Copernican theory he was eventually tried by the inquisition, a medieval system established by Pope Gregory IX in

1233 and lasting until the nineteenth century, for the investigation of heresy. He was found guilty in 1633 and sentenced to a life imprisonment that was subsequently commuted to permanent house arrest.

It was in *On Motion and on Mechanics* that Galileo clearly identified four elements to be considered in the study of machines.

> [T]he first is the weight to be transferred from one place to another; second is the force or power that must move it; third is the distance between the beginning and the end of the motion; and fourth is the time which the change must be made – which time comes to the same thing as the swiftness and speed of the motion. (Drabin and Drake, 1960)

Among other things, Galileo demonstrated that the velocity of a falling body increased in proportion to the time from the start of its fall regardless of its weight. The concept, that the product of a force and a distance over which it was moved was conserved, was apparent in a letter written in 1637 to Huygens by Descartes (1596–1650):

> The invention of all machines is founded upon nothing but a single principle, which is that the same force which is able for example to lift 100 pounds to a height of two feet, is also able to lift 200 pounds to a height of one foot, or a weight of 400 pounds to a height of one-half foot, etc. (Hiebert, 1962)

Christiaan Huygens (1629–95) was born in The Hague to a prominent family that served the Dutch royalty. He studied law and mathematics at Leiden and at Breda. In the 1650s and 1660s he traveled to Paris where he met Pascal and Roberval and to London where he interacted with Wallis and became a member of the Royal Society. This was also his most productive period during which he completed research in mathematics, mechanics, astronomy and optics. He invented the pendulum clock in 1656.

Newton combined and augmented these and other concepts to form the basis of what is now known as classical mechanics. His ideas are to be found in the definitions, laws, propositions, and corollaries that he set down about 1665 but did not publish until 1687 in *Principia*. Although they have been superseded by Einstein's theories they are still sufficiently accurate at the scales we shall consider.

From Book I

Definition II: The quantity of motion is the measure of the same, arising from the velocity and quantity of matter conjointly.

Law I: Every body continues in its state of rest, or of uniform motion in a right line, unless it is compelled to change that state by forces impressed upon it.

Law II: The change of motion is proportional to the motive force impressed; and it is made in the direction of the right line in which that force is impressed.

Law III: To every action there is always opposed an equal reaction; or, the mutual actions of two bodies upon each other are always equal, and directed to the contrary parts.

Proposition IV. Cor VI: . . . the centripetal forces will be inversely as the squares of the radii;

From Book III

Proposition VII. Theorem VII: There is a power of gravity pertaining to all bodies, proportional to the several quantities of matter which they contain.

As we shall see below, calculus is an ideal tool for describing and applying these concepts. However, despite being the foremost developer of this branch of mathematics Newton did not use it in an explicit manner to describe his laws. According to Straub (1981) Daniel "Bernoulli was the first to link Newton's ideas with Leibniz's calculus which he learned from his brother Nicholas."

Daniel Bernoulli (1700–82) was the son of Johann (see Section 2.5) . His early studies were in philosophy and logic and then in medicine. However his 1724 publication of *Exercitationes mathematicæ* led to his appointment to the St.Petersburg Academy (1725–33) where he worked with Euler after 1727. During that period Daniel developed his principle work on hydrodynamics, a term that he introduced in print in 1738 in *Hydrodynamica*.

4.2 Base Units

Mechanics is a subject that deals with entities such as force, momentum, and energy. These are terms which are used loosely in everyday language but which have clearly defined meanings in physics and associated disciplines. They are constructed from the three basic elements of length, mass, and time, which in turn have been precisely described within the International System of Units, internationally abbreviated to "SI" (AMS, 1974). They are given below (see appendix C for conversions).

Length, symbols x,y,z in Cartesian coordinates, **r** as a position vector: The meter (m) is the length equal to 1,650,763.73 wavelengths in vacuum of the radiation corresponding to the transition between the levels 2 p_{10} and 5 d_5 of the krypton-86 atom.

Mass, symbol M: The kilogram (kg) is the unit of mass; it is equal to the mass of the international prototype of the kilogram. (The international prototype of the kilogram is a particular cylinder of platinum–iridium alloy which is preserved in a vault at Sèvres, France, by the International Bureau of Weights and Measures.)

Time, symbol t: The second (s) is the duration of 9,192,631,770 periods of the radiation corresponding to the transition between the two hyperfine levels of the ground state of the cesium-133 atom.

4.3 Derived Units

From the fundamental base units the various quantities of mechanics may be derived. James Maxwell (1881) followed this procedure.

4.3.1 Speed

This is familiar to everyone as the rate with respect to time at which a length is traveled. In calculus it becomes

$$u = \frac{dx}{dt},$$

or in vector notation,

$$\mathbf{v} = \frac{d\mathbf{r}}{dt}. \tag{4.1}$$

The units are m s^{-1}.

4.3.2 Acceleration

The rate at which speed changes with respect to time is given by

$$\frac{du}{dt} = \frac{d^2x}{dt^2},$$

or

$$\frac{d\mathbf{V}}{dt} = \frac{d^2\mathbf{r}}{dt^2}. \tag{4.2}$$

The units are m s^{-2}.

4.3.3 Force

Force was first fully defined by Newton in his Second Law, the change of motion is proportional to force. From his Definition II we see that "motion" refers to the product of mass and velocity, which we now call *momentum*. He does not state that the change is considered with respect to time although his subsequent work indicates that it is. Also, the proportionality constant is "1" so that the Law is described today as the change of momentum with respect to time is equal to force,

$$\frac{d(M\mathbf{V})}{dt} = \mathbf{F}. \tag{4.3}$$

The units are kg m s^{-2}, which is also designated to be a newton, or abbreviated to N. By convention in SI, when units take on a person's name the first letter is in lower case but when abbreviated the upper case is used.

Because in most situations, mass remains constant, it may be taken outside the differential and we have the popular statement of the Second Law that the product of mass and acceleration is equal to force,

$$M\frac{d\mathbf{V}}{dt} = M\frac{d^2\mathbf{r}}{dt^2} = \mathbf{F}. \tag{4.4}$$

Another way of thinking about momentum results from integrating Equation (4.4) with respect to time. Then, if force is constant,

$$M\mathbf{V}_2 - M\mathbf{V}_1 = \mathbf{F}(t_2 - t_1), \tag{4.5}$$

the change in momentum is equal to the time effect of the force, called the *impulse*.

Curved Motion

Of much interest in the seventeenth century was the motion of planets. Their orbits were being observed much more accurately and people sought to explain them. One of those was Newton, who succeeded in applying his postulates concerning motion and gravity. For circular motion there had to be acceleration toward the center of curvature called centripetal acceleration. Today it is usually described in the following way. In figure 4.1 the circular motion of a particle at P, which is located at some angle, θ, to an origin line OA, may be described in terms of either the linear velocity, v, or the angular velocity, ω, and the radius of curvature, r (see cross product in section 2.6.2),

$$\text{Angular velocity} = \omega = \frac{d\theta}{dt}, \tag{4.6}$$

$$\text{Tangential velocity} = v = r\omega = r\frac{d\theta}{dt}. \tag{4.7}$$

In order to find the accelerations of P along the tangential and normal directions for constant radius of curvature and constant angular velocity we consider

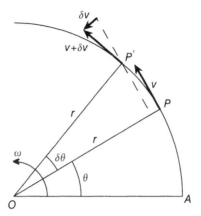

Figure 4.1 Curved motion

a small change in the position of P to P' through angle $\delta\theta$ as shown in figure 4.1. Note that δv also subtends an angle $\delta\theta$ at P'. As a result

$$\delta v = (v + \delta v)\cos \delta\theta - v. \qquad (4.8)$$

But from the series expansion of cosines and sines for x in radians [Equations (2.91) and (2.92)],

$$\cos(x) = 1 - \frac{x^2}{2!} + \frac{x^4}{4!} - \cdots,$$

$$\sin(x) = x - \frac{x^3}{3!} + \frac{x^5}{5!} - \cdots,$$

we see that for very small x, $\cos(x) = 1$, and $\sin(x) = x$. Therefore, the acceleration along the tangent is given by

$$\text{Tangential acceleration at P} = \lim_{\delta t \to 0}\left(\frac{\delta v}{\delta t}\right) = \frac{dv}{dt} \qquad (4.9)$$

which, for constant angular velocity, is zero.

Similarly the change in velocity along the normal PO is $(v+\delta v)\sin\theta - 0 = (v + \delta v)\delta\theta$. From this we may obtain the acceleration along PO, known as the centripetal acceleration,

$$\text{centripetal acceleration at P} = \lim_{\delta t \to 0}\left(\frac{v\,\delta\theta + \delta v\,\delta\theta}{\delta t}\right) = v\frac{d\theta}{dt} \qquad (4.10)$$

Substitution from Equation (4.7) gives alternative forms of the centripetal acceleration [also see Equation (2.51)] where in vector notation this is $\omega \times \mathbf{r}$,

$$v\frac{d\theta}{dt} = v\omega = r\left(\frac{d\theta}{dt}\right)^2 = r\omega^2 = \frac{v^2}{r}. \tag{4.11}$$

For curved motion this centripetal acceleration is necessary. Otherwise, according to Newton's Second Law the object at P would continue along a tangent i.e. along v.

According to the Third Law, the centripetal acceleration must be balanced by an equal and opposite action. This we sense as a force pointing away from the center of curvature such as the tension in a string holding a circulating object. This is often called the centrifugal force $Mr\omega^2$.

The whole derivation could have been done in vector notation. Here we shall generalize, allowing for varying radius of curvature, r, and varying angular velocity, $(d\theta/dt)$, i.e. angular acceleration.

$$\mathbf{V} = r\frac{d\theta}{dt}\mathbf{i}_\tau + \frac{dr}{dt}\mathbf{i}_n \tag{4.12}$$

where $\mathbf{i}_n$ is the unit vector along OP, the unit normal vector, and $\mathbf{i}_\tau$ is the unit tangent vector pointing counterclockwise. The differential of a unit tangent is the rate of turning of the unit tangent (see Section 2.2), so it points along PO, i.e. negative $\mathbf{i}_n$.

$$\frac{d\mathbf{i}_\tau}{dt} = -\frac{d\theta}{dt}\mathbf{i}_n. \tag{4.13}$$

Similarly the differential of a unit normal is the rate of turning of the unit normal and points in the $\mathbf{i}_\tau$ direction,

$$\frac{d\mathbf{i}_n}{dt} = \frac{d\theta}{dt}\mathbf{i}_\tau. \tag{4.14}$$

Therefore, differentiation of Equation (4.12) with respect to time to obtain acceleration gives

$$\begin{aligned}
\frac{d\mathbf{V}}{dt} &= \frac{dr}{dt}\frac{d\theta}{dt}\mathbf{i}_\tau + r\frac{d^2\theta}{dt^2}\mathbf{i}_\tau + r\frac{d\theta}{dt}\frac{d\mathbf{i}_\tau}{dt} + \frac{d^2r}{dt^2}\mathbf{i}_n + \frac{dr}{dt}\frac{d\mathbf{i}_n}{dt} \\
&= \frac{dr}{dt}\frac{d\theta}{dt}\mathbf{i}_\tau + r\frac{d^2\theta}{dt^2}\mathbf{i}_\tau + r\left(\frac{d\theta}{dt}\right)^2\mathbf{i}_n + \frac{d^2r}{dt^2}\mathbf{i}_n + \frac{dr}{dt}\frac{d\theta}{dt}\mathbf{i}_\tau \\
&= \left(2\frac{dr}{dt}\frac{d\theta}{dt} + r\frac{d^2\theta}{dt^2}\right)\mathbf{i}_\tau - \left(r\left(\frac{d\theta}{dt}\right)^2 - \frac{d^2r}{dt^2}\right)\mathbf{i}_n. \tag{4.15}
\end{aligned}$$

The first two terms associated with $\mathbf{i}_\tau$ give the counterclockwise tangential acceleration and the second two associated with $\mathbf{i}_n$ give the acceleration towards the center of curvature. The negative sign in the last term, d^2r/dt^2, indicates that the radius of curvature is assumed to be increasing. If r is constant then this reduces to Equations (4.9) and (4.11).

Another way of writing Equation (4.15) is

$$\frac{d\mathbf{V}}{dt} = (2\dot{r}\dot{\theta} + r\ddot{\theta})\mathbf{i}_r - (r\dot{\theta}^2 - \ddot{r})\mathbf{i}_n. \qquad (4.16)$$

Also, note that

$$(2\dot{r}\dot{\theta} + r\ddot{\theta}) = \frac{1}{r}\frac{d(r^2\dot{\theta})}{dt}. \qquad (4.17)$$

Because a radian has no units it will be seen that angular acceleration has the same units as linear acceleration, m s^{-2}.

Angular Momentum

Angular momentum is defined as the product of the linear momentum and the radius of curvature. For a mass, M, at radius, r, (located at P in figure 4.1) this is

$$Mur = M\omega r^2. \qquad (4.18)$$

The units are kg m^2 s^{-1}. Note that this contains an additional length unit compared to linear momentum.

Gravity and Weight

Newton did not write down in *Principia* the complete equation for the force of gravity quite in the form we use it today (Newton, 1687, Appendix Note 54, p. 670), yet it is clear that he was fully aware of it. This is apparent from the extracts given at the beginning of the chapter and also from Book III Proposition VII. Theorem VII (pp. 414–15):

> *That there is a power of gravity pertaining to all bodies, proportional to the several quantities of matter which they contain.*

> Cor. II. The force of gravity towards the several equal particles is inversely as the square of the distance of places from the particles; as appears from Cor. III, Prop. LXXIV, Book I.

Today we state that the force of gravity between two objects is proportional to the product of their masses and inversely proportional to the square of the

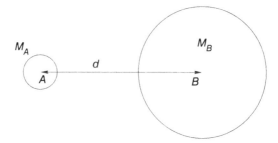

Figure 4.2 Force of gravity between two objects M_A and M_B

distance between their centers of gravity,

$$\mathbf{F}_G = G \frac{M_A M_B}{d^2}, \tag{4.19}$$

(see figure 4.2). The proportionality constant G has a magnitude of $6.67 \times 10^{-11} \mathrm{N\,m^2\,kg^{-2}}$. A short book which deals with gravity in fairly simple terms is by Phillips (1968). It is remarkable that at the end of the twentieth century while most physical constants are known to several decimal digits G, called "Big G," is only known to three (Kestenbaum, 1998).

For a particular object with mass M_A, interacting with the earth, which has a mass $M_B = 5.975 \times 10^{24}$ kg, the force of gravity is affected only by the distance squared, d^2. Hence we may write, from Equation (4.19) for two different distances r and $r + z$,

$$\mathbf{F}_{G,r} r^2 = \mathbf{F}_{G,r+z}(r + z)^2 = G M_A M_B,$$

or, for unit mass,

$$\frac{\mathbf{F}_{G,r} r^2}{M_B} = \frac{\mathbf{F}_{G,r+z}(r + z)^2}{M_B} = G M_A. \tag{4.20}$$

The radius of the earth varies with $r_{equator} = 6378.1$ km and $r_{pole} = 6356.9$ km. Let f_0 in m s^{-2} be the average $\mathbf{F}_{G,r}/M_B$ at sea level, then

$$\mathbf{f}_z = \mathbf{f}_0 \frac{r^2}{(r + z)^2} = \mathbf{f}_0 \left(1 + \frac{z}{r}\right)^2$$

$$= \mathbf{f}_0 \left(1 - 2\frac{z}{r} + \cdots\right)$$

$$\approx \mathbf{f}_0 \left(1 - 2\frac{z}{r}\right)$$

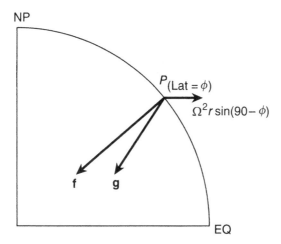

Figure 4.3 True gravity and the rotation effect

With the average magnitude of r, $r_0 = 6371$ km,

$$\mathbf{f}_z = \mathbf{f}_0(1 - 0.314 \times 10^{-6}z). \tag{4.21}$$

Earth rotation produces a centripetal acceleration [Equation (4.11)] which is difficult to separate from $\mathbf{f}$ so it is usually combined with it (figure 4.3). The centripetal effect accounts for the bulge of the earth at the equator. It is only of the order of 0.03m s^{-2} at a maximum so the standard acceleration of gravity, $\mathbf{g}$, is taken to be 9.80665 m s^{-2}. The earth rotation effect means that "horizontal" surfaces, which are defined as perpendicular to the direction of $\mathbf{g}$, are not spherical surfaces. Also, since a constant g is used even into the lower stratosphere, adjustments are made to meteorological height to compensate for that [see Equation (7.47) below].

The important distinction between *mass* and *weight* is clearly evident from the equation for weight, $\mathbf{F}_g$, which is now written as the product of mass and the acceleration of gravity,

$$\mathbf{F}_g = M\mathbf{g}. \tag{4.22}$$

Too often we confuse the two as we do in everyday language where we tend to use the two words interchangeably.

4.3.4 Quantity of Motion

For a long time it was unclear how the concept of motion should be defined and how it related to force. People "worked" and moved objects which produced motion and other changes. Various practical systems like levers and pulleys had been developed that apparently allowed the effect to be multiplied (Lindsay, 1975). The Greeks had recognized that in a pulley system the speed at which the rope was pulled was much greater than the speed at which a weight was raised. As indicated at the beginning of this chapter these concepts were further refined by Galileo and Descartes. It also seemed evident that within these processes something was conserved. To Descartes this quantity was related to speed:

> For while it is true that motion is only the behavior of matter which is moved, there is, for all that, a quantity of it which never increases nor diminishes, although there is sometimes more and sometimes less of it in some of its parts; it is for this reason that when a part moves twice as rapidly as another part, and this other part is twice as great as the first part, we have a right to think that there is as much motion in the smaller body as in the larger, and that every time and by as much as the motion of one part diminishes that of some other part increases in proportion. (Descartes, 1644)

In other words, the quantity of motion that was conserved, was the same phrase that Newton later used to define momentum.

A different view was held by Leibniz. In 1686 in *A Brief Demonstration of the Memorable Error of Descartes and Others Concerning the Natural Law According to Which They Claim that the Same Quantity of Motion Is Always Conserved by God, a Law That They Use Incorrectly in Mechanical Problems* published in *Acta Eruditorum* he argued that *vis motrix*, his term for momentum, was not conserved. Nine years later, also in *Acta Eruditorum*, Leibniz introduced two new terms *vis mortua* (dead force) being potential force for motion and *vis viva* (living force), the actual force of motion. His examples showed that vis viva was proportional to the product of mass and the square of the velocity, a quantity that had been used previously in 1673 by Huygens.

Over the next century there was a debate over the best definition of "conserved force" and the concept of vis viva. Adherents to the use of the square of the velocity were Johann and Daniel Bernoulli and Leonard Euler. In 1803 Carnot (1803) was still using vis viva for MV^2 and vis mortua for weight.

4.3.5 Energy

The word "energy" is of Greek origin where $\grave{\epsilon}\nu\epsilon\rho\gamma\acute{\eta}\varsigma$ referred to force or vigor of expression (Oxford, 1978). It had also been used in a more relevant sense by Henry More in 1642 "Platonicall Song of the Soul" who wrote "as the light of the Sunne is the energie of the Sunne." Whereas the germs of the various forms of energy were present in the literature the separate terms to describe them took a long time to emerge.

Work

Descartes (see quotes at the beginning of the chapter) had set down the principles of mechanical work in 1637 but no formula had appeared. With his interest in conservation and quantity of motion Johann Bernoulli investigated a number of ideas relating of the equilibrium of forces. Through these studies he appears to have been the first to have defined work, which he called "energy," in terms of force and distance. In a letter to Pierre Varignon in 1717 he introduced the terms *vîtesse virtuelles* (virtual velocity) for the movement of forces and *energie* to refer to force times distance (Hiebert, 1962, p. 83). The following is Varignon's account of the correspondence:

Conceive [says he] several different forces which are acting along different lines or directions of tendency to maintain in equilibrium a point, a line, a surface, or a body; conceive also that we impress on the whole system of forces a small displacement, either parallel to itself along any direction, or about any fixed point; it is easy to see that by this displacement each of these forces will advance or recede in its direction, unless some one or more of the forces have their directions perpendicular to the direction of the small displacement; in which case the force or the forces will neither advance nor recede: for these advancements or recessions, which I call virtual velocities, are nothing other than the amounts by which each line of tendency increases F or decreases because of the small displacement; and these increments or decrements are found by drawing a perpendicular from the end of each line of tendency which will cut off from the line of tendency of each force in the neighboring position, to which it has been brought by the small displacement, a small portion which will be the measure of the virtual velocity of this force. For example, let P (figure 4.4) be a point in the system of forces which is in equilibrium; let F be one of these forces, which pushes or pulls the point P in the direction FP or PF, let Pp be a small straight line which the point P describes in a small displacement, by which the line of tendency FP takes the position fp, which will either be exactly parallel to FP, if the displacement of the system is so made that all its points move parallel to a given straight line; or, prolonged, will make with *F P* an infinitely small angle, if the displacement of the system takes place around a fixed point. Now draw PC perpendicular to fp, and you will have Cp representing the virtual velocity of the force F, so that F × Cp is what I call the energy. Notice that Cp is either positive or negative with respect to other similar lines: it is positive, if the point P is pushed by the force F and the angle FPp is obtuse; and negative if the angle FPp is acute; but on the contrary, if the point P is pulled by the force, Cp is negative, if the angle FPp is obtuse, and positive, if it is acute. All this

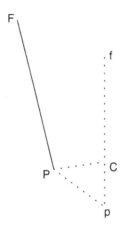

Figure 4.4 Varignon's figure from Bernoulli

being well understood, I lay down [says M. Bernoulli] the following

General Proposition, Theorem XL

> In every case of equilibrium of forces, in whatever way they are applied, and in whatever directions they act on one another, either mediately or immediately, the sum of the positive energies will be equal to the sum of the negative energies, taken as positive.

In later papers Johann elaborated on the idea.

Carnot (1803) "evaluated the force of a man, horse, or the like . . . [by examining] . . . the work it is capable of doing in a given time." His son, Sadi called work "motive power" (Carnot, 1824).

The term *work* to denote this form of mechanical energy did not become popular until later. Gaspard Gustave de Coriolis (1792–1843) used it to describe *mechanical work* (travail méchanique) in connection with vis viva. He was a Parisian, trained at École Polytechnique and employed there from 1816 on the recommendation of Cauchy. In 1829 he became chair of mechanics at the École Centrale des Arts et Manufactures and in 1832 moved to the École des Ponts et Chaussés to work with Navier in applied mechanics. His name is now associated with the acceleration term in the equation of motion that appears in the rotating frame of reference.

In his discussion of heat in 1834 Clapeyron used the term *mechanical action* but Clausius replaced this by *work* in 1850 (see section 5.4 below).

The Scot, William John Macquorn Rankine (1820–72), who followed in his father's footsteps as a railroad engineer before he became a professor at the University of Glasgow in 1855, contributed much to the understanding of energy. In 1855 he wrote:

> "Work" is the variation of an accident by an effort, and is a term comprehending all phenomena in which physical change takes place. Quantity of work is measured by the product of the variation of the passive accident by the magnitude of the effort, when this is constant; or by the integral of the effort, with respect to the passive accident, when the effort is variable.
> Let x denote a passive accident;
> X an effort tending to vary it;
> W the work performed in increasing x from x_0 to x_1: Then
>
> $$W = \int_{x_0}^{x_1} X\, dx, \quad \text{and}$$
>
> $$W = X(x_1 - x_0), \quad \text{if } X \text{ is constant.}$$
>
> Work is represented geometrically by the area of a curve, whereof the abscissa represents the passive accident, and the ordinate, the effort. (Rankine, 1855, 216–17)

(The term "accident" refers to an occurrence or realization of a variable.)

A very clear statement in current terms appeared first in 1881 on page 5 of volume I of James Clerk Maxwell's treatise on electromagnetism:

> The unit of Work is the work done by the unit of force acting through the unit of length measured in its own direction. Its dimensions are $[M\, L^2\, T^{-2}]$. (Maxwell, 1881)

Therefore, even though the basic principles of "work" were set down by Bernoulli in 1717 it took another 150 years before our current wording emerged.

In vector notation the definition of work is:

$$\text{Work} = WE = \mathbf{F} \cdot \mathbf{r} = Fr\cos\theta, \tag{4.23}$$

where θ is the angle between the force and the direction of movement r. The units are $\text{kg m}^2\,\text{s}^{-2}$.

Potential Energy

Again the idea of potential energy is evident early in, for example, Leibniz's vis mortua. In 1724 Johann Bernoulli, in *Discourse on the Laws of the Communication of Motion*, elaborated on the vis mortua of weight and of the spring. (Lindsay, 1975)

Lazare-Nicholas-Marguerite Carnot(1753–1823) in a book in 1803 recognized the relationship between vis mortua, equivalent to a weight being raised through a distance, and vis viva, equivalent to the mass and velocity squared. Therefore he, like Jean Le Rond d'Alembert (1743), saw the debate over which "vis" was a measure of the force of a body, was a matter of semantics.

Euler called the integral of force with respect to distance *effort* while Lagrange called it *potential* (Boissonnade and Vagliente, 1997, p. xxxviii). It was not until 1857, however, that William Rankine coined the term *potential energy*. Potential energy is nothing more than another expression for latent work, vis mortua. A spring is compressed by work, whereby the potential energy resides in the spring. Similarly, an object requires work to be raised above some standard level, again providing a stored energy. The components of the latter are mass, acceleration of gravity and height,

$$\text{Potential Energy} = PE = M \times \mathbf{g} \times z, \qquad (4.24)$$

where z is height above some standard base, often taken to be mean sea level. The units are, as before, $\text{kg m}^2 \text{ s}^{-2}$.

Kinetic Energy

That most of the concepts of mechanical energy were current in scientific circles by the end of the eighteenth century is evident from the lectures presented by Thomas Young. Young (1773–1829) was a precocious child learning to read at two and becoming proficient in several languages and science by the time he was a teenager. He became a physician but soon turned to natural philosophy with the support of Benjamin Thompson. He prepared popular lectures for the Royal Society in 1802 and 1803. In them he dealt with vis viva and introduced the term energy for it:

> The term energy may be applied, with great propriety, to the product of mass or weight of a body, into the square of the number expressing its velocity. Thus, if the weight of one ounce moves with a velocity of a foot in a second, we call its energy 1; if a second body of two ounces have a velocity of three feet in a second, its energy will be twice the square of three, or 18. (Young, 1807)

His use of "energy" was ignored by his contemporary scientific colleagues.

Four years later Lagrange (1811) used calculus to produce the conservation of energy equation and to show that a factor of two was involved in the relationship between potential energy and vis viva. Joseph Louis Lagrange (1736–1813) was born in Turin as Lodovico Giuseppe Lagragia but adopted the French name early in life. His father wanted him to become a lawyer but he soon gravitated to mathematics and science. As a teenager he developed an analytical method in the calculus of variations with which he impressed Euler, the founder of the field. Lagrange was appointed professor of geometry at 20 and worked in Turin until 1766 when he moved to the employ of the King of Prussia in Berlin. In 1787, the year after Frederick the Great died, Lagrange moved to Paris where he later became a professor at the École Polytechnique. Lagrange made important contributions to calculus, mechanics, and probability theory.

The term *kinetic energy* was introduced by Coriolis in 1829. It is defined by

$$\text{Kinetic Energy} = KE = \frac{1}{2}MV^2, \qquad (4.25)$$

which, like work and potential energy, also has units of kg m^2 s^{-2}.

Conservation of Energy

Following Lagrange we may write the equation for force (Equation 4.4) from Newton's Second Law as

$$M\frac{dV}{dt} = M\frac{d^2\mathbf{r}}{dt^2} = \mathbf{F}.$$

In order to obtain the total effect of this force while it operates on the object we need to add up its continuous action along the path that it takes. One way to do this is to rewrite $Md\mathbf{V}/dt$ to include the path increment, $\delta\mathbf{r}$, in derivative form,

$$M\frac{d\mathbf{V}}{d\mathbf{r}}\frac{d\mathbf{r}}{dt} = M\frac{d^2\mathbf{r}}{dt^2} = \mathbf{F}.$$

All products of vectors here are assumed to be "dot" products. Now integrate from "a" to "b" with respect to the position vector $\mathbf{r}$,

$$\int_a^b M\mathbf{V}\,d\mathbf{V} = \int_a^b M\frac{d^2\mathbf{r}}{dt^2}d\mathbf{r} = \int_a^b \mathbf{F}\,d\mathbf{r},$$

which becomes

$$\left[\frac{1}{2}MV^2\right]_a^b = \left[\frac{1}{2}M\left(\frac{dr}{dt}\right)^2\right]_a^b = \left[Fr\cos\theta\right]_a^b. \qquad (4.26)$$

The elements involving M are two equivalent expressions for the change or difference in kinetic energy between points "a" and "b" along the path, and the element involving $\mathbf{F}$ is the change or difference in potential energy between the same points, i.e.

$$KE_b - KE_a = -(PE_b - PE_a).$$

The first negative sign on the right is necessary because F points downwards and $\mathbf{r}$ points upwards: $\theta = \pi$. Put in another way, if these are rearranged, we can say that the sum of kinetic and potential energies remain the same along the path: the sum is conserved,

$$KE_a + PE_a = KE_b + PE_b. \tag{4.27}$$

In order to illustrate this conservation relationship we may consider the energy involved in raising an object from the floor and letting it fall back again. Let the object have a mass of 3 kg and the height lifted be 2 m. Then the force, weight, is $3\,\text{kg} \times 9.8\,\text{ms}^{-2} = 29.4\,\text{kg m s}^2$. The work performed to raise this weight will be $29.4\,\text{kg m s}^2 \times 2\,\text{m} = 58.8\,\text{kg m}^2\,\text{s}^{-2}$ which is also the potential energy of the object, say at "a." If the object is now allowed to fall from "a" the starting potential energy, PE_a, is $58.8\,\text{kg m}^2\,\text{s}^{-2}$ and the starting kinetic energy, KE_a, is zero. At the floor, b, the height is zero so PE_b is zero. Therefore, with these magnitudes in Equation (4.27),

$$0 + 58.8\,\text{kg m}^2\,\text{s}^{-2} = KE_b + 0,$$

and so, as the object hits the floor,

$$\frac{1}{2}MV^2 = 58.8\,\text{kg m}^2\,\text{s}^{-2}.$$

From this, substituting for M, we may find the magnitude of the velocity as the object hits the floor,

$$V = \sqrt{\left(2 \times \frac{58.8\,\text{kg m}^2\,\text{s}^{-2}}{3\,\text{kg}}\right)} = 6.26\,\text{m s}^{-1}.$$

Of course, the conversion of work to potential energy is not efficient and we have ignored friction with the air, so the calculations are only an approximation of the process. Also, there are other forms of energy yet to be defined. We shall deal with these issues after we have considered gases and how energy affects them.

4.3.6 Power

Power is the time rate of energy change,

$$\text{Power} = \frac{dE}{dt}, \tag{4.28}$$

in units of $\text{kg m}^2\,\text{s}^{-2}\,\text{s}^{-1} = \text{watt} = \text{W}$. As an illustration a 40 W light bulb uses $40\,\text{kg m}^2\,\text{s}^{-2}$ of energy per second. The light that you see is not equal to that amount because it is distributed spherically and some of the energy is in non-visible wavelengths and some is transferred by conduction. Many electrical utility companies charge by the unit of a kilowatt-hour. That is in some ways a strange unit to use because

$$1 \, \text{kW-hr} = 10^3 \, \text{W-hr} = 10^3 \, \text{kg m}^2 \, \text{s}^{-2} \, \text{s}^{-1} \times 60 \times 60 \, \text{s}$$
$$= 3,600,000 \, \text{kg m}^2 \, \text{s}^{-2}.$$

Another way of writing the conservation of energy is to set Power equal to zero,

$$\frac{dE}{dt} = 0. \tag{4.29}$$

Similarly, another way of writing the rate of working, as will be used in section 5.5.2, from Equation (4.23), is

$$\frac{d(WE)}{dt} = \mathbf{F} \cdot \frac{d\mathbf{r}}{dt} = \mathbf{F} \cdot \mathbf{v}, \tag{4.30}$$

where $\mathbf{F}$ is assumed not to change with time.

4.3.7 Force and Momentum of Curved Motion for Solid Bodies

So far we have considered only the linear momentum equation based upon Newton's Second Law, Equation (4.3). As we move to extended bodies we must take into account other factors, not least of which is rotational motion of the object itself. This is in addition to curved motion of the center of gravity which, for say a planet, is distinguished from rotation by the term *revolution*.

Basic to consideration of rotation is the concept of the turning moment. The first known recorded analysis of this everyday experience is from Archimedes (ca. 250 BC). He pointed out that the effect of a force on an extended body such as a light rod depended upon the distance of the force from the pivot point as well as the magnitude of the force itself. In it's simplest form this is the law of the lever which states that the forces on a light pivoted rod will be balanced if their turning moments (*force × distance*) are balanced (figure 4.5), i.e.

$$Ax_A + Bx_B = Cx_C \tag{4.31}$$

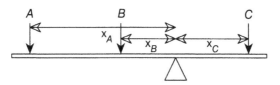

Figure 4.5 A balanced light bar with weights A, B, and C at distances x_A, x_B, and x_C respectively from the pivot point

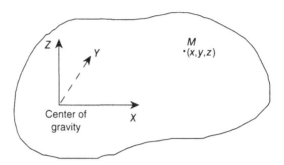

Figure 4.6 Balance point for a solid, the center of gravity

This brings us to the idea of the *center of gravity*, the point around which the effect of the *long range* forces (e.g. gravity and the effect of earth rotation), as opposed to *short range* forces (e.g. pressure and friction) cancel out. In the situation where gravity is the only existing long range force, this is the balance pivot point: the point around which the turning moments cancel. In figure 4.6 for the xz plane

$$\sum Mgx + \sum Mgz = g\left(\sum Mx + \sum Mz\right). \qquad (4.32)$$

In the general case, if Mg is constant throughout the body,

$$\sum (\mathbf{i}x + \mathbf{j}y + \mathbf{k}z) = 0. \qquad (4.33)$$

All long range forces may be considered as operating only on the center of gravity. Furthermore, if these are the only ones which are operating, the body may be replaced by a *mass point*: all the mass is located at the center of gravity. This procedure significantly simplifies problems involving long range forces.

Rotation is different from linear motion in that it requires two forces, which are parallel, equal, and opposite, known as a *couple* or a *torque*, to alter it. Usually we tend to think of only one force because we can speed up or slow down a wheel with a single hand. What we forget is that the wheel is subject to other forces and that in concert with the additional force a couple is produced. For example, the tangential force imparted by the hand is counteracted by a parallel, equal and opposite force from the axle, or if the wheel is unattached, by the tangential force at the ground (figure 4.7).

Note that because couples are parallel, equal, and opposite there can be no resultant force, only a turning moment. Also, the turning moment of a couple

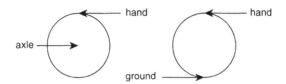

Figure 4.7 Torque produced by the hand on a wheel. In the diagram on the left the wheel is assumed to be supported only by the axle and the torque is set up with a force equal and opposite to the hand by the axle. In the diagram on the right the wheel is supported by the ground and the equal and opposite force is produced at the ground

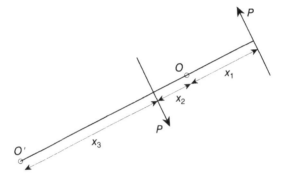

Figure 4.8 The torque of "P" couple about O and O'

is constant about any point (see figure 4.8). The turning moment for couple P

$$\text{from } O = P(x_1 + x_2)$$
$$\text{from } O' = P(x_1 + x_2 + x_3) - Px_3$$
$$= P(x_1 + x_2)$$

For a solid body, as shown in figure 4.6 with $r^2 = x^2 + y^2$, the angular momentum is the sum of the angular momenta of the individual elements (Equation 4.18):

$$\sum M\omega r^2 = \omega \sum M r^2 = \omega I. \qquad (4.34)$$

I, in units of kg m^2, is known as the *moment of inertia*. The units of solid body angular momentum are the same as those for individual mass points, $\text{kg m}^2 \text{ s}^{-1}$.

From this discussion it is also clear that angular velocity of a solid object in a given plane may be calculated from the difference between any two parallel velocities divided by the perpendicular distance separation. For the xy plane this may be written

$$\omega = \frac{\delta v}{\delta x} \qquad (4.35)$$

As examples of particular solids, we find that the angular momentum of a sphere of uniform density is

$$I_s \omega = \frac{2}{5} M_{sp} r^2 \omega, \qquad (4.36)$$

and, for a spherical shell of uniform density, it is

$$I_h \omega = \frac{2}{5} M_{sh} \frac{(r_1^5 - r_2^5)}{(r_1^3 - r_2^3)} \omega. \qquad (4.37)$$

For a very thin shell of average radius r Equation (4.36) reduces to

$$I_h \omega = \frac{2}{3} M_{sh} r^2 \omega. \qquad (4.38)$$

As implied above, angular momentum for a solid body may be changed only through the action of a torque. Indeed, Lagrange (1811, p. 182) considered it to be a constant:

the sum of the products of the masses with their velocities and with the perpendiculars from the paths of the bodies to the center is a constant quantity.

Thus the rate of change of angular momentum, Newton's Second Law applied to curved motion, is

$$I \frac{d\omega}{dt} = I \frac{d^2\theta}{dt^2} = \sum (Px) = T \qquad (4.39)$$

where T is the torque, the sum of the turning moments on the body. The units are $kg\,m^2\,s^{-2}$, the same as kinetic energy.

Consequently if the torques are reduced to a minimum, angular momentum tends to be conserved. The principle is used in friction toys where the energy is built up in a flywheel. It is the same principle which is inherent in the gyroscopic compass.

Another way of looking at angular momentum comes from integrating Equation (4.39) with respect to time as we did for Equation (4.5). Then, when T is constant,

$$I \omega_2 - I \omega_1 = T (t_2 - t_1). \qquad (4.40)$$

This says that the change in angular momentum is due to the time effect of the torque which is called the *moment of impulse*.

Kinetic Energy of Rotation

The kinetic energy equation for rotation may be derived in a similar way to linear kinetic energy as in Equation (4.26). In this case the rate of change of angular momentum equation (Equation 4.39) is integrated with respect to angle, θ,

$$\int I \frac{d\omega}{d\theta} \frac{d\theta}{dt} \, d\theta = \int T \, d\theta. \tag{4.41}$$

Therefore

$$\frac{1}{2} I \omega_2^2 - \frac{1}{2} I \omega_1^2 = T(\theta_2 - \theta_1) \tag{4.42}$$

which states that the change of kinetic energy is equal to the work done by the external forces. The units are $\text{kg m}^2 \text{ s}^{-2}$.

4.4 Discussion

Like the mathematics included in chapter 2, the material on mechanics is usually assumed knowledge at the beginning of a course on atmospheric dynamics. As already indicated, it is *classical mechanics* that was developed prior to the twentieth century, resting as it does upon Kepler's and Newton's Laws. It turns out to be a special case of the laws of relativity but one which has been extremely successful in describing interactions between forces and motions in most of our everyday experiences. It is sufficient for calculating the planetary orbits and variations that are needed in investigations of solar radiation receipt. It is also quite adequate for looking at relationships in the atmosphere. It is unfortunate, therefore, that there are few courses today that are devoted to, or even emphasize, classical mechanics.

An attempt has been made in this chapter to review the simple concepts of mechanics starting from the elementary units of mass, length, and time in order to introduce the basic equations. These will be used in chapter 7 in the discussion of relationships used in fluid dynamics.

Chapter 5

Thermodynamics

5.1 Definitions

Thermodynamics is the study of heat and temperature. Sometimes its scope is restricted to the macroscopic behavior of matter. At other times, as considered here, it includes microscopic behavior called *kinetic theory*, *statistical mechanics*, and in a more fundamental form *quantum mechanics*. Historically the macroscopic characteristics were investigated first so these will be discussed first in the following sections.

The overall entities with which thermodynamics deal, such as a parcel of air or the earth as a whole, are usually referred to as *systems*. Often, as with an air parcel, they have no clearly defined boundaries or size and we typically refer to them in terms of a unit mass or unit volume. We characterize, or in other words, describe, the state of these systems by variables of state such as pressure, volume, and temperature. When variables are normalized by mass or volume the adjective *specific* is prepended. Thus volume per unit mass becomes *specific volume*. The normalization factor becomes obvious from the units employed. We may consider these systems as being theoretically *isolated*, that is, *closed*, or in contrast *open* in various ways from their environment. Thus thermal isolation is termed *adiabatic* whereas thermally open is *diathermic*.

Note that V and the lower case v in thermodynamics refer to volumes. Previously and in future sections V and v refer to velocity.

5.2 The Equation of State – the Macroscopic Approach

At the same time that forces and motions of solid objects were being defined and related in a precise way the state variables of gases were also being investigated.

Their inter-relationships are described by an equation known as the *equation of state*.

5.2.1 Temperature

Although Galileo had demonstrated the variation of temperature through the effect of expanding and contracting air in a glass flask probably in the 1590s, there is no evidence that he added a scale to it (Magie, 1963). This development is usually credited to Santorio Santorio (1561–1636), an Italian known as Sanctorius who reported on an air thermometer in his *Commentaria in artem medicinalem Galeni* in 1612 (Grmek, 1981). Numerous individuals refined the instrument in the succeeding century. Liquid in glass was introduced about 1630. Robert Hooke put a scale on his instrument with a single fixed point, the freezing point of water in 1665 (Middleton, 1969), and at about the same time Huygens suggested that either the freezing or the boiling points might be used. Carlo Renaldini was probably the first to use the two changes in the state of water as fixed points on a temperature scale in 1694 (Roller, 1950). In an anonymous article in the *Philosophical Transactions of the Royal Society* that Magie (1963) attributes to Newton, the writer had several marked points on his instrument including the melting and boiling points of water and the melting point of lead (Newton, 1701). Ole or Olaus Christensen Römer (1644–1710) wrote up a detailed set of instructions on how to produce a calibrated mercury thermometer. However, such an instrument is usually attributed to Fahrenheit. Daniel Gabriel Fahrenheit (1686–1736), born in Danzig but who lived much of his life in England and Holland, where he visited Römer, began to make calibrated mercury in glass thermometers about 1716 and wrote about them in 1724 (Fahrenheit, 1724). For his fixed points he used a mixture of ice, water, and sea-salt (sal-ammoniac) for 0 and mouth or armpit for 96 (Middleton, 1969, p. 62).

The Swedish astronomer, Anders Celsius (1701–44), reported in 1742 on his thermometer which used 0 for boiling and 100 for freezing. It is unclear who inverted the scale during the next few years (Middleton, 1969).

A tacit assumption made in dealing with the temperature of objects is that *when two systems are at the same temperature as a third, they are at the same temperature as each other*. This statement is sometimes called the *Zeroth Law of thermodynamics* (Sears, 1953).

5.2.2 Pressure

The Italian mathematician and physicist, Evangelista Torricelli (1608–47), who was secretary to Galileo and succeeded him as professor in Florence, invented the barometer in about 1643. He described his experiment in a letter in 1644 (see translation in Middleton, 1969). In 1648 Pascal's brother-in-law, Florin Perier, made measurements near the foot and at the top of the Puy-de-Dôme in France and demonstrated that atmospheric pressure decreased with height.

5.2.3 Pressure and Volume – Boyle's or Mariotte's Law

In 1657 Guericke invented an air pump that was improved by Robert Hooke. This instrument was used by Boyle to relate pressure and volume. Robert Boyle (1627–91), born at Lismore Castle in Ireland, was the fourteenth child of an affluent family. He was well educated in England (Eton) and in Geneva. He was influenced by Bacon and Descartes and studied with Wallis in Oxford. Settling in London in 1668 he was one of the founders of the Royal Society. In 1660 he published *New Experiments Physico-Mechanical, Touching the Spring of the Air and its Effects* (Boyle, 1662a). In it he reported on a number of discoveries using the air pump. One, now known as *Boyle's Law*, showed that pressure and volume were inversely related. Symbolically this may be written

$$pV = \text{constant}, \tag{5.1}$$

or as

$$pV = p_0V_0. \tag{5.2}$$

Boyle also supported the hypothesis, following Bacon and Descartes, that matter was composed of very small particles.

In his *Nature del'Air* published in 1679 Mariotte stated the same relationship without reference. As a result his claim to discovering the law is often disputed. Nevertheless it is usually known as *Mariotte's Law* on the continent of Europe. Edme Mariotte has unknown birth records but died in 1684 being honored as the man who introduced experimental physics into France. He had a central role in the Paris Academy of Science from its formation in 1666 until his death.

5.2.4 Gay-Lussac's Law

In 1787 the French physicist, chemist, inventor, and balloonist, Jacques Alexander César Charles, is reported to have discovered that, at constant pressure, volume was proportional to temperature. However, he published nothing, and it was Gay-Lussac who credited him with the relationship. Joseph Louis Gay-Lussac (1778–1850), like his father added "Lussac" from the family property in the Limoges to his name to distinguish himself from the other Gay's in the region. He attended the École Polytechnique where he also later became a professor of physics (1808) and of chemistry (1809). In 1801–02 at the encouragement of Berthollet and Laplace he carried out very detailed experiments to resolve the currently conflicting evidence about the expansive properties of gases.

> he concluded that equal volumes of all gases expanded equally with the same increase in temperature. Over the range of temperature from 0°C. to 100°C. the expansion was 1/266.66 of the volume at 0°C. for each rise in temperature. Similar research was carried out by Dalton at about the same time. Dalton's work, however, was considerably less accurate. (Crosland, 1981b)

As an equation this could be expressed as

$$V = V_0(1 + aT) \tag{5.3}$$

where $a = 1/266.66$ and T was temperature.

5.2.5 Dalton's Law

John Dalton (1766–1844), born into a Quaker family in Cumberland, England, was trained at a school in Kendal. After eight years as a professor at a college in Manchester he set up he own academy in 1800. He worked in mathematics, mechanics, chemistry, and meteorology and lectured throughout Britain. In 1822 he visited Paris where he met many of the important scientists of the time such as Laplace, Berthollet, and Gay-Lussac. Besides his findings with regard to volume expansion at constant pressure he discovered that total atmospheric pressure was the sum of the pressures of the separate gases, *Dalton's Law of partial pressures*. This finding, first published in 1793 in his *Meteorological Observations* (Thackray, 1981), may be written:

$$p = p_1 + p_2 + p_3 + \cdots \tag{5.4}$$

5.2.6 Avogadro's Hypothesis

Of importance in the development of the current form of the equation of state is *Avogadro's Hypothesis* published in 1811. Amedeo Avogadro (1776–1856) was born in Turin where he spent much of his life as an academic. His well-known hypothesis claimed that the number of *molecules* in a gas was related to volume:

> That gases always unite in a very simple proportion by volume, and that when the result of the union is a gas, its volume also is very simply related to those of its components. But the quantitative proportions of substances in compounds seem only to depend on the relative number of molecules which combine, and on the number of composite molecules which result. It must then be admitted that very simple relations also exist between the volumes of gaseous substances and the numbers of simple or compound molecules which form them. The first hypothesis to present itself in this connection, and apparently even the only admissible one, is the supposition that *the number of integral molecules in any gas is always the same for equal volumes, or always proportional to the volumes* ... The hypothesis we have just proposed is based on that simplicity of relation between the volumes of gases on combination, which would appear to be otherwise inexplicable. (Crosland, 1981a)

As a base, the lightest atom, the hydrogen atom, was used and given the value of one atomic mass unit (amu), generally called the molecular weight. Thus for hydrogen, "m," the symbol used here for molecular weight, is set equal to 1. Today carbon 16 has been adopted for the base. It should be noted that the use of the word "weight" is a poor one since it is really the "mass" that is being referenced. The connection between the molecular "weight" and the mass of a gas was made using the concept of the *mole* which became popular during the nineteenth century. The mole, or as used today, the kilogram mole (or kg-mole or kmol), is defined as the mass of a gas in kilograms that numerically is the same as its molecular weight. Therefore the number of moles, n, in a gas of mass M is given by

$$n = \frac{M}{m} \text{ kmol.} \tag{5.5}$$

From his hypothesis, even though there was no clear distinction between atoms and molecules, Avogadro was able to estimate the relative masses of molecules. As stated by Parkinson (1985):

> Among the consequences of accepting the Avogadro Hypothesis is the determination that water is composed of twice as many molecules as oxygen molecules since, upon being separated into the two gases, the volume of the resulting hydrogen is twice that of the resulting oxygen. The masses of hydrogen and

oxygen in water being in the ratio $1:8$, the conclusion is therefore that the molecular weight of oxygen is 16 times that of hydrogen.

Surprisingly his discovery was essentially ignored until 1860 when the Italian, Stanislao Canizzaro (1826–1910), used the hypothesis to calculate the relative molecular and atomic weights of several gases.

It was not until about 1908 that the actual number of molecules in a specified volume at the same temperature and pressure, N_0, known as *Avogadro's Number* was determined accurately. This was done by Perrin in his experiments with Brownian motion. Jean Baptiste Perrin (1870–1942) was a professor at the Sorbonne from 1897 to 1940. He was very much involved in with the institutional development of science in France but he had to move to the United States in 1940 because of his anti-fascism (Stuewer, 1981).

The current estimate for N_0 under the conditions of standard temperature and pressure (273.16 K, 101.325 kPa) is $6.022169 \times 10^{26} \, \text{kmol}^{-1}$. This gives the volume of an ideal gas as $22.4136 \, \text{m}^3 \, \text{kmol}^{-1}$.

5.2.7 Equation of State

With Gay-Lussac's work all the necessary ingredients were available at the beginning of the nineteenth century for linking the three variables of state, p, V, and T. Several researchers including Poisson were cognizant of it. Therefore by 1824 Carnot was able to develop the equation in a footnote to a memoir in the following form (Carnot, 1824),

$$p = N \frac{(T + 267)}{V}. \tag{5.6}$$

Nicolas Leonard Sadi Carnot (1796–1832) was the son of a French public servant who gave up that career to concentrate on science and the education of his sons. Lazare-Nicholas-Marguerite Carnot (1753–1826), the father, had appointed Napoleon to his first command and was an eminent scientist in his own right. As a physicist, engineer, and mathematician he had used vis viva, Mv^2, as the basis of his investigation of machines as mentioned in section 4.3.4. Needless to say, Sadi was trained well and was accepted into the elite École Polytechnique in Paris at age 16. He studied analysis, mechanics, descriptive geometry, and chemistry. His teachers included Poisson, Guy-Lussac, and Ampère. Sadi served intermittently in the military but continued to conduct research especially as it related to steam engines.

Today the magnitude of the number added to T in Equation (5.6) is, of course, 273.16 to convert celsius to kelvin. Carnot called the quantity N "the elastic force of this same air at constant volume 1, but at the temperature zero" divided by 267. If pV in Equation (5.3) is substituted into Boyle's law (5.2) we see that it is equal to $p_0 V_0 a$.

Carnot's relationship was restated ten years later by Clapeyron who used the letter "R" instead of "N", defined as $p_0V_0/(267 + T_0)$ (Clapeyron, 1834). Benoit-Pierre-Émile Clapeyron (1799–1864) also attended École Polytechnique, graduating in 1818, four years after Carnot. In 1850 Clausius (1850) similarly adopted "R" of which he wrote, "This last constant is in so far different for the different gases that it is inversely proportional to their specific gravities." The equation now read

$$pV = RT, \tag{5.7}$$

where $R = (p_0V_0)/(273 + T_0)$. By then experiments had shown $1/a$ to be 273. He estimated R for air with the following substitutions

$$R = \frac{10333 \, \text{kg m}^{-2} \times 0.7733 \, \text{m}^3 \, \text{kg}^{-1}}{273 \, \text{K}} = 29.26 \, \text{m K}^{-1}. \tag{5.8}$$

If we convert to today's units by multiplying by the acceleration of gravity (necessary to convert kilograms mass to weight) we obtain approximately the correct magnitude for the gas constant for dry air, R_d, of $286.7 \, \text{J kg}^{-1} \, \text{K}^{-1}$. Rudolf Clausius (1822–88) was born in Köslin, Prussia. He entered the University of Berlin in 1840 and graduated with a doctorate from Halle in 1847. He taught at the Royal Artillery School in Berlin before becoming a professor in Zurich (1855–67) partly as a result of his 1850 paper.

Later, Equation (5.7) was normalized by division by n kmol. Thus we have the current equation of state:

$$pv = R^*T; \qquad pV = nR^*T \tag{5.9}$$

where p is in Pa, v in $\text{m}^3 \, \text{kmol}^{-1}$, and T in kelvin. R^* is the universal gas constant and has a magnitude of $8314.34 \, \text{J kmol}^{-1} \, \text{K}^{-1}$.

Equation (5.9) is an idealized relationship and other equations relating the state variables are possible. Real gases only approximate it especially over limited ranges of the variables so its full name is the *equation of state for an ideal gas*. Fortunately empirical observations show that Equation (5.9) is a very good approximation to the range of variables with which we deal in the earth's atmosphere.

In thermodynamics much use is made of graphs. This started with James Watt (1736–1819), a professor at the University of Glasgow, who is famous for his work on steam engines and who has been recognized by having the unit of the rate of energy flux named after him. He used what have been called *indicator diagrams*. The equation of state is usually plotted on p–v axes as in figure 5.1.

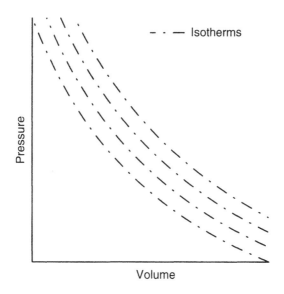

Figure 5.1 Indicator diagram displaying temperature from the equation of state for an ideal gas on p–v coordinates

Typically for atmospheric applications, n in Equation (5.9) is replaced from Equation (5.5). Then

$$p\frac{\mathcal{V}}{M} = \frac{R^*}{m}T.$$

Since m is available for different gases, e.g. $m_d = 28.9\,\text{kg kmol}^{-1}$ for dry air and $\mathcal{V}/M = \rho$, density, the ideal equation of state for dry air may be written

$$p_d = \rho_d R_d T \qquad (5.10)$$

where $R_d = R^*/m_d = 287\,\text{J kg}^{-1}\,\text{K}^{-1}$, the gas constant for dry air. Tables 1–6 in (Iribarne and Godson, 1973) shows that the dry air in the troposphere deviates from Equation (5.10) by less than 0.2 percent.

5.2.8 The Equation of State – the Microscopic Approach

The idea that matter was composed of small particles had its origin in Greek thought and it appears in the scientific literature throughout the sixteenth to eighteenth centuries. Indeed, Daniel Bernoulli in chapter 10 of *Hydrodynamica* (1738) had sketched out a *kinetic gas theory* (Straub, 1981). However, the concept was not well developed until the 1860s.

In that decade, following the early work of Clausius, both Maxwell and Boltzmann were independently developing the kinetic theory of gases. It applies Newtonian mechanics to individual molecules as if they were perfectly elastic spheres which, of course, they are not, but then makes this approach more reasonable by extrapolating to a large number of molecules using probability theory.

The basic assumptions are that:

1 A volume of gas contains a large number of molecules, Avogadro's Number, N_0. The number is approximately 6×10^{26} kmol^{-1} $\approx 3 \times 10^{25}$ m^{-3} = 3×10^{16} mm^{-3}.

2 The distance between the molecules is much larger than their diameters. From 1, each molecule occupies a cube of $1/(3 \times 10^{25})$ m^3 $\approx 30 \times 10^{-27}$ m^3 or 3×10^{-9} m on a side (i.e. apart). The diameter of a molecule is estimated to be $\approx 3 \times 10^{-10}$ m. Therefore, molecules in a gas are about 10 times their diameter apart.

3 The only forces present occur at collisions. Otherwise they move in straight lines.

4 The collisions, between molecules and between molecules and walls are perfectly elastic. Also, the walls are perfectly smooth so there is no change in the speed parallel to the wall and only the sign changes normal to the wall.

5 The molecules are uniformly distributed throughout the gas.

6 The motion is random.

7 The speed of a molecule varies between zero and the speed of light. At sea level the mean speed is about 400 m s^{-1}. Their mean free path (average distance between collisions) is approximately 7×10^{-8} m, so their rate of collision is in the order of 10^{33} m^{-3} s^{-1}. The latter large number indicates how rapidly characteristics in a gas are transmitted and local equilibrium is maintained.

Applying these assumptions we may arrive at a number of relationships that will be given on the following pages between the macro and micro variables (Sears, 1953).

The equation of state was originally derived by Clausius in 1857 (Daub, 1981):

$$pV = \frac{1}{3}NM\,\bar{v}^2 \tag{5.11}$$

where N is the number of molecules per unit volume, M is the mass of the molecule and $\bar{v}$ is the average velocity of the molecules. Because n, the number of moles, is the number of molecules divided by Avogadro's Number,

$$n = \frac{N}{N_0},$$

and since Boltzmann's constant, k, is defined by

$$k = \frac{R^*}{N_0}, \tag{5.12}$$

we may derive

$$\frac{1}{2} M \bar{v}^2 = \frac{3}{2} kT. \tag{5.13}$$

This is the origin of the popular description of temperature that it is proportional to the mean translational kinetic energy of the molecules.

Molecular theory has shown that the energy is distributed equally between different gases and between the different coordinate directions. In addition, there are exchanges between the other forms of molecular energy. This process is known as *the equipartition of energy*. It is through this mechanism that radiation, which is not directly connected to the translational energy of the molecule, produces temperature changes (see section 6.2).

5.3 Atmospheric Composition

A full decomposition of the gases of the atmosphere was not performed until the twentieth century. Several reasons are given for this late accomplishment, the most obvious being that gases are generally invisible. Indeed, the term *gas*, introduced by Johannes Baptista (1579–1644) of Brussels, may have derived from the German word for *ghost* although Bohren and Albrecht (1998) point to the letter "g" as coming from the Greek *chaos*. Also, in the seventeenth and eighteenth centuries there was considerable confusion between "air" and "flame." Nevertheless, Robert Boyle anticipated many later discoveries with his statement from *Memoirs for a General History of the Air*:

> The Schools teach the air to be a warm and moist element, and consequently a simple and homogeneous body. Many modern philosophers have, indeed, justly given up this elementary purity of air, yet few seem to think it a body so greatly compounded as it really appears to be. The atmosphere, they allow, is not absolutely pure, but with them it differs from true and simple air only as turbid water from clear. Our atmosphere, in my opinion, consists not wholly of purer aether or subtile matter which is diffused thro' the universe, but in great numberless exhalations of the terraqueous globe; and the various materials

that go to compose it, with perhaps some substantial emanations from celestial bodies, make up together, not a bare indetermined feculancy, but a confused aggregate of different effluvia.

. . .

I conjecture that the atmospherical air consists of three different kinds of corpuscles: the first, those numberless particles which, in the form of vapours or dry exhalations, ascend from the earth, water, minerals, vegetables, animals, etc.; in a word, whatever substances are elevated by celestial of subterraneal heat, and thence diffused into the atmosphere. The second may be yet more subtle, and consist of those exceedingly minute atoms the magnetical effluvia of the earth, with other innumerable particles sent out from the bodies of celestial luminaries, and causing, by their impulse, the idea of light in us. The third sort is its characteristic and essential property, I mean permanently elastic parts. (Ramsay, 1896)

A contemporary of Boyle, John Mayow (1645–79) who entered Wadham College, Oxford, at 16, graduated with a Law degree, but practiced medicine, wrote two books on respiration that were reported in the *Philosophical Transactions* (Mayow, 1668). In them he stated that he considered the atmosphere to consist of two kinds of gases: one, "aëial nitre particles," which was necessary for the support of life; and the rest, which was incapable of either.

A few years later George Ernest Stahl (1660–94) a professor at the University of Halle introduced the term *phlogiston*. This was a substance that was thought to be separated from materials during combustion. What was left after oxygen and carbon dioxide were removed was "phlogistic material." Lavoisier denied its existence in 1775 and it was largely abandoned as a concept by 1800. During the intervening period, however, it played a central role in the development of ideas concerning the composition of the atmosphere. Antoine-Laurent Lavoisier (1743–94) was a public servant, social reformist, and general scientist. As a geologist he mapped northern France and was the first to recognize that geological time had been marked by a series of epochs that could be identified by retreating and advancing sea level. Despite that work he is, of course, most acclaimed for his genius in chemistry (Guerlac, 1981). After he was guillotined Lagrange commented, "It took them only an instant to cut off that head, and a hundred years may not produce another like it." Besides the overthrow of the phlogiston theory Lavoisier was pivotal in showing that the atmosphere was composed of different gases. Following Scheele in Sweden and Priestley in England he demonstrated that oxygen was an important component. By 1781 Henry Cavendish had established that air, freed from carbon dioxide by potash was composed 71.16 percent of phlogisticated air (nitrogen) and 20.84 percent of dephlogisticated air (oxygen) very closed to current estimates.

Table 5.1 Composition of the atmosphere (permanent gases)

Gas	Formula	Molec. frac. %	Molec. wt.
Dry air		100.0	28.966
Nitrogen	N_2	78.09	28.016
Oxygen	O_2	20.95	32.000
Argon	Ar	0.93	39.944
Neon	Ne	0.0018	20.183
Helium	He	0.000524	4.003
Krypton	Kr	0.0001	83.7
Hydrogen	H_2	0.00005	2.016
Xenon	Xe	0.000008	131.3

Source: List (1958)

Table 5.2 Composition of the atmosphere (variable gases)

Gas	Formula	Molec. frac. %	Molec. wt.
Water vapor	H_2O	0 to 4	18.016
Carbon dioxide	CO_2	0.035	44.010
Methane	CH_4	0.00017	
Nitrous oxide	N_2O	0.00003	
Ozone	O_3	0.000004	48.00
Dust particles		0.000001	
Chloroflurocarbons	CFCs	0.000000001	

Argon, as a component was reported by Lord Rayleigh and William Ramsay in 1895. Tables 5.1 and 5.2 list recent estimates of the composition of the atmosphere.

5.4 Heat

5.4.1 General Comments on Temperature and Heat

As we have seen in section 5.2.8 temperature is proportional to the mean translational kinetic energy of the molecules. As such it represents only part of the total energy of a substance. However, it is a characteristic that is directly observable by a person's senses and through changes in the properties such as volume (e.g. length of mercury) or state (e.g. solid to liquid, etc.).

Heat is a term that everyone uses and is often confused with temperature. In fact, heat can be a confusing concept and indeed, as described in the following sections, it took scientists several centuries to explain. One of the main reasons for confusion is that heat is a form of energy flow or conversion and not a form that can be contained in an object. This is similar to work. When a person physically *works* food energy is converted into mechanical energy. However, it is not possible the define how much work is contained in a person. The authors Bohren and Albrecht (1998) even advocate that, "We try to use the word *heat* as little as possible . . . " They prefer the terms *heating* and *working*. Unfortunately, both *heat* and *work* are used so frequently in the meteorological literature that a change is very unlikely to occur.

In meteorology we may recognize that heat flow, as a form of energy flow, occurs through molecular contact (conduction), through mass transfer (turbulent exchange of water or air parcels in the vertical and horizontal) and through radiation. We also take into account that the mass transfer of heat involves the transfer of moisture (latent energy) as well as the more obvious temperature (sensible energy).

Typically there is a net heat flow between objects as a result of different temperatures. Thus the collision of molecules between the finger tips and an object of higher temperature leads to a transfer of kinetic energy to molecules in the finger that the brain recognizes as "warm" or "hot." As shown below, in solids and liquids heat flow leads to temperature change, and in a gas it leads to both a temperature and a pressure change. On the other hand, we also find that the temperature in a gas may change as a result of pressure change and without a heat flow.

5.4.2 Early Development of the Concept of Heat

In the sixteenth century heat was considered variously as an unusual form of substance or as a function of motion. Lindsay (1975) used Pierre Gassendi (1592–1655) and Francis Bacon (1561–1626) as respective proponents of these approaches.

Robert Boyle (see section 5.2.3) even described the process of conversion from motion:

> EXPERIMENT VI – When, for example, a smith does hastily hammer a nail, or such like piece of iron, the hammered metal will grow exceeding hot. (Boyle, 1662a)

It was the material theory, however, which took hold, especially after Lavoisier introduced the term *caloric* to describe it (Lavoisier, 1789). This hypothesized an "elastic fluid." The concept was used throughout the nineteenth century by researchers studying heat, including Sadi Carnot and in his early papers by William Thompson. Hot objects were believed to contain more caloric than cold objects and heat transfer was accomplished by the flow of caloric.

It was already recognized that, given a flow of heat, different substances displayed different amounts of temperature change. This characteristic was named *specific heat*. Also, it was apparent that quantities of heat were absorbed during changes in phase. The proportionality constant or characteristic that connected the two is known as *latent heat*. In the posthumous publication of his lectures in 1803 Joseph Black (1728–99), a physician and professor of chemistry at Glasgow and Edinburgh, there is a discussion of both of these characteristics. In a number of experiments Black demonstrated that mixing equal amounts of water and quicksilver at different temperatures lead him to conclude:

> Quicksilver, therefore, has less *capacity* for the matter of heat than water (if I may be allowed to use this expression) has; it requires a smaller quantity of it to raise its temperature by the same number of degrees. (Magie, 1963)

Also,

> Black describes an experiment, in which, when equal masses of water and ice were exposed to similar sources of heat, it is found that the temperature of the water rose regularly, while that of the water formed from the melting ice did not rise. (Magie, 1963)

It is not clear when these discoveries were made but both concepts existed and estimates of these constants were being made by the late eighteenth century.

The opposing view to heat being a substance was held by Benjamin Thompson. Thompson (1753–1814), born in Massachusetts, was a loyalist in the Revolution. He joined the British army where he rose to colonel and was knighted by George III. He later moved to Bavaria where he headed the army and was made a count of the Holy Roman Empire in 1793. He took the name Count Rumford after the old name of Concord, New Hampshire (Brown, 1981). As a soldier he became interested in the boring of cannons and serendipity (Ford, 1968). He established the Royal Institution in London in 1799 for the education of the underprivileged and the Academy of Arts and Sciences in Munich in 1801. Among other scientific work, he performed experiments demonstrating that friction could generate heat without changing the nature of the material supplying the heat and that heat had no perceptible effect on the mass of a material object. With regard to caloric he stated:

> It is hardly necessary to add that anything which any insulated body, or system of bodies, can continue to furnish without limitation cannot possibly be a material substance; and it appears to me to be extremely difficult, if not quite impossible, to form any distinct idea of anything, capable of being excited and communicated, in the manner the heat was excited and communicated in these experiments, except it be motion. (Rumford, 1798)

Despite having this work printed simultaneously in six different publications Count Rumford's findings were not widely acknowledged.

The first calculation of the numerical relationship between mechanical energy and heat was performed by von Mayer. Julius Robert von Mayer (1814–78) was a German physician and physicist. He was expelled from the University of Tübingen medical program in 1837 but passed the state exams in 1838. In 1842 he published *On the Forces of Inorganic Nature* in which he gave a magnitude for energy conversion in modern units of 3,580 J cal^{-1} (von Mayer, 1842). (The current magnitude is 4,190.0.) His work received little attention at the time although Clausius (1850) was aware of it and pointed out that in 1845 C. Holtzman "calculates the numerical value of the constant a in the same way as Mayer had already done, and obtains a number which corresponds with the heat equivalent obtained by Joule in entirely different ways." Holtzman called "the unit of heat which by its entrance into a gas can do the mechanical work a – that is, to use definite units, which can lift a kilograms through 1 meter." This is the unit of energy we now call a joule $= J = kg\,m^2\,s^{-2}$.

James Prescott Joule (1818–89) was the son of a wealthy brewer in Salford, England. He was educated at home with one of his teachers being John Dalton. Most of his life was spent in Manchester as an amateur scientist. The first presentation of his experimental findings on the mechanical equivalence of heat were given in a lecture in a church reading room and published in a newspaper, the *Manchester Courier* on May 5 and 12 1847 (Joule, 1847a).

Another presentation was given that June to the British Association in Oxford.

> The author exhibited and described an apparatus, consisting of a brass paddle-wheel working in a vessel filled with liquid, with which he had repeated the experiments brought before the Cambridge Meeting of the Association. By these experiments he had shown that heat is invariably produced by the friction of fluids in exact proportion to the force expended. Two series of experiments had been made – one on the friction of water, the other on the friction of sperm-oil. In the former of these series the heat capable of raising the temperature of a pound of water 1° was found to be equal to the mechanical force capable of raising a weight of 781.5 pounds to the height of one foot; whilst in the series of experiments on the friction of sperm-oil, the same quantity of heat was found to be equal to a mechanical force represented by 782.1 pounds through one foot. (Joule, 1847b)

(These numbers translate into 4201.9 and 4205.1, very close to the current 4190.0 J, the average energy required to raise the temperature of 1 kilogram 1 kelvin.) The presentation excited at least one person in the audience, the 22 year old William Thompson (destined to be Lord Kelvin) who later used

Joule to perform experiments that required high accuracy. Unfortunately Joule's lack of advanced mathematical training prevented him keeping up with the new scientific advances of the late eighteenth century.

Thus, by 1847 a fairly precise connection had been made between heat and mechanical energy. Also, it was generally accepted that matter was composed of small particles, atoms and molecules. In the words of Clausius (1850): "heat is not a substance, but consists in motion of the least parts of bodies." The connection between these motions and energy had to await the development of statistical mechanics.

5.5 The First Law of Thermodynamics

5.5.1 Clausius's Statement

This is a statement in thermodynamics of the conservation of energy which, as we have seen in the last section, is an old concept. It was the basis of Carnot's 1824 paper but it was not clearly stated until Clausius published his book in 1875 (Clausius, 1875). Iribarne and Godson (1973), like Clausius, called it *The First Principle* but others identified it as *The First Law*.

Clausius defined his First Principle in his Equation (II) on page 27 of his *Mechanical Theory of Heat* (Clausius, 1875) as

$$dQ = dH + dJ + dW \qquad (5.14)$$

where dQ is "an indefinitely small quantity of heat ... imparted," dH is "the indefinitely small increment" of "the total heat existing in the body, or more briefly Quantity of Heat," dJ is the increment of Ergal and dW is the increment of work. Previously Clausius had defined Ergal on page 11:

> Hamilton gave to it the special name of "force function;" ... In the later and more extended investigations with regard to quantities which are expressed by this function, it has become needful to introduce a special name for the *negative* value of the function, or in other words for that quantity, the *subtraction* of which gives the work performed; and Rankine proposed for this the term "potential energy." This name sets forth very clearly the character of the quantity; but it is somewhat long, and the author has ventured to propose in its place the term "Ergal." (Clausius, 1875)

J is, in fact, equal to $-p\mathcal{V}$.

Clausius (1850) also defined the sum of H and J in 1850 as *internal energy*, U:

$$U = H + J = H - p\mathcal{V}. \qquad (5.15)$$

This is also Maxwell's relation for *enthalpy*,

$$H = U + p\mathcal{V}, \tag{5.16}$$

a quantity which is usually introduced in textbooks as occurring "frequently in thermodynamic equations" without explanation (Sears, 1953). It is the sum of the internal and potential energies. For an air column in hydrostatic equilibrium extending to the top of the atmosphere these remain in constant proportion to one another (Haurwitz, 1941).

The internal energy may be shown to be only a function of temperature so isotherms in indicator diagrams are also isopleths of constant U. This term also defines the form of energy associated with temperature in a gas, often referred to in error as "heat." It is one of four distinct forms of energy considered to be resident in the atmosphere, the others being potential, kinetic, and latent energies.

Substituting the differential Equation (5.16) into Equation (5.14) Clausius arrived at his Equation (III) on page 27:

$$dQ = dU + dW \tag{5.17}$$

which is the general form of the First Law occurring in textbooks at the present time. Note that they usually define dW with the opposite sign from Clausius.

5.5.2 Recent Statement

Like heat, as mentioned in section 5.4, a system cannot contain a quantified amount of work. The energy exchange due to "working" is defined by an area on a p–V indicator diagram yet there are an infinite number of paths that a system may follow in changing from one equilibrium state to another. Consequently increments of heat and work cannot be written as exact differentials and they cannot be integrated in the normal way (see end of section 2.4). To signify this, authors often use special symbols for increments of heat and work such as δQ and δW, so Equation (5.17) becomes

$$\delta Q = dU + \delta W. \tag{5.18}$$

To avoid this difficulty Bohren and Albrecht divide Equation (5.18) through by dt and set $\delta Q/dt = \mathcal{Q}$ and $\delta W/dt = \mathcal{W}$ and rearrange to obtain their Equation (3.1) (Bohren and Albrecht, 1998):

$$\frac{dU}{dt} = \mathcal{Q} + \mathcal{W}. \tag{5.19}$$

The energy involved in "working" is associated with the interaction of pressure and volume through expansion and contraction. Thus, since in mechanics "rate of working = force × speed" [Equation (4.30)], if we consider an expansion such that the area of the surface of the expanding gas remains a constant, A,

$$W = -pA \frac{dx}{dt} = -p \frac{d\mathcal{V}}{dt}. \tag{5.20}$$

The negative sign indicates that expansion subtracts energy from the system.

5.5.3 Specific Heat Capacities and the Gas Constant

The specific heat capacity (per unit mass) at constant volume, c_v, may be defined as

$$c_v = \frac{du}{dT} \text{ J kg}^{-1}\text{ K}^{-1}, \tag{5.21}$$

where lower case letters represent "specific" quantities, i.e. "per unit mass." Integration over appropriate limits gives us a measure of the internal energy in a gas

$$u = c_v T = IE \text{ J kg}^{-1}. \tag{5.22}$$

This shows, as indicated in section 5.5.1, that the internal energy is a function only of temperature.

Similarly the specific heat capacity (per unit mass) at constant pressure may be defined as

$$c_p = \frac{dh}{dT} \text{ J kg}^{-1}\text{ K}^{-1}. \tag{5.23}$$

Again, integration over appropriate limits gives us a measure of the enthalpy in a gas

$$h = c_p T \text{ Jkg}^{-1}. \tag{5.24}$$

This shows that the enthalpy is also a function only of temperature and may be represented by the isotherms on the indicator diagram.

The specific heats also are related to the gas constant,

$$c_p - c_v = R = \frac{R^*}{m}. \tag{5.25}$$

Since these are all "constants" the equation applies generally. So, for dry air $c_v = 718$, $R_d = 287$, and $c_p = 1005 \text{ J kg}^{-1}\text{ K}^{-1}$. The molecular weight of dry air is $m_d = 28.964 \text{ kg kmol}^{-1}$. The variation in c_p over a range of atmospheric temperatures and pressures is provided by Iribarne and Godson. These are given in SI units in table 5.3.

Table 5.3 Specific heat at constant pressure for dry air

p, kPa	$-80°C$	$-40°C$	$0°C$	$+40°C$
0	1002.3	1003.7	1004.0	1005.7
30	1004.4	1004.0	1004.4	1006.1
70	1006.5	1005.3	1005.3	1006.5
100	1009.0	1006.5	1006.1	1007.3

5.5.4 Standard Forms of the First Law

The relationships between heating and the gas characteristics may now be written without derivation in the forms usually found in meteorological publications:

$$Q_M = c_v \frac{dT}{dt} + p \frac{d\alpha}{dt}, \qquad (5.26)$$

and

$$Q_M = c_p \frac{dT}{dt} - \alpha \frac{dp}{dt}, \qquad (5.27)$$

where Q_M is the heating (energy per unit time) per unit mass and $\alpha = 1/\rho$.

5.6 The Carnot Cycle

We have already referred to Sadi Carnot's insightful papers. Because they lay the ground work for the *Second Law* we shall summarize them now. Sadi's thesis was based upon three assumptions: 1) that perpetual motion was impossible; 2) that heat was conserved and that the amount absorbed or released during a process depended only on the initial and final states; and, 3) that motive power could be produced whenever a temperature difference existed between two reservoirs. In 1823 he published *Recherche d'une formule propre à representer la puissance motrice de la vapeur d'eau*, in which he sought a mathematical expression for work produced by one kilogram of steam. He used a three stage engine. In the following year in *Reflexions sur la puissance motrice du feu et sur les machines propres à développer cotte puissance* he theorized an ideal four stage engine, now known as the *Carnot Engine*, and the processes involved, the *Carnot Cycle*. The paper attracted little attention but ten years later Émile Clapeyron elaborated on the theory and clarified it by using indicator diagrams.

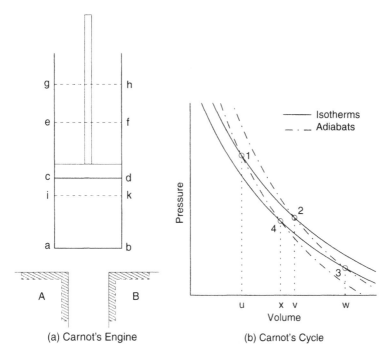

(a) Carnot's Engine   (b) Carnot's Cycle

Figure 5.2 Carnot's Engine and Cycle (a) Fig. 1 from Carnot's 1824 paper. (b) Clapeyron's Fig. 1 describing the sequence of events in (a)

The sequence of events as described by Carnot, with reference to Clapeyron's diagrams, is as follows (Mendoza, 1977):

In figure 5.2(a) we have two bodies A and B each kept at constant temperature with A having a higher temperature than B. The object labeled with lower case letters is a closed (at the bottom) cylinder containing a movable piston.

(1) The bottom of the cylinder a-b is placed in thermal contact with A at a-b so energy is allowed to flow freely between them. When the temperature of the air in the cylinder is the same as in A the position of the piston is at cd. On the p–v diagram in (b) this corresponds to a position between 1 and 2, actually at volume x.

(2) The piston gradually rises and takes the position ef. The body A is all the time in contact with the air, which is thus kept at a constant temperature during rarefaction. The body A furnishes the caloric necessary to keep the temperature constant. [We have now reached point 2 in (b) having followed an isotherm.]

(3) The body A is removed, and the air [in the cylinder] is then no longer in contact with any body capable of furnishing it with caloric. The piston

meanwhile continues to move, and passes from the position ef to the position gh. The air is rarefied without receiving caloric, and its temperature falls. Let us imagine that it falls thus till it becomes equal to that of the body B; at this instant the piston stops, remaining at the position gh. [We have now reached point 3 on (b) along the dash-dot line, which we now call an *adiabat*, but which was unnamed at that time.]

(4) The air is placed in contact with the body B; it is compressed by the return of the piston as it is moved from the position gh to the position cd. This air remains, however at constant temperature because of its contact with the body B, to which it yields its caloric. [We are now at position 4.]

(5) The body B is removed, and the compression is continued, which being then isolated, its temperature rises. The compression is continued till the air acquires the temperature of body A. The piston passes during this time from the position cd to the position ik. [We are now at position 1.]

(6) The air is again placed in contact with the body A. The piston returns from the position ik to the position ef; the temperature remains unchanged. [This brings us again to (2).]

(7) The step described under number (3) is renewed, then successively the steps (4), (5), (6), (3), (4), (5), (6), (3), (4), (5); and so on.

[So we have Carnot's Cycle, a sequence of isothermal and adiabatic processes that returns the system to its original state.]

5.7 Dry Adiabats and Potential Temperature

In 1834 Clapeyron had no name for the change of temperature during processes that occurred under thermally isolated conditions. However, in 1875 Clausius wrote:

In this case the relation between pressure and volume is given by Equation (45), viz.:

$$\frac{p}{p_1} = \left(\frac{v_1}{v}\right)^k .$$

The curve of pressure corresponding to this equation (Fig. 7) [p–v indicator diagram of an adiabat] falls more steeply than that delineated in Fig. 5 [p–v indicator diagram of an isotherm]. Rankine has given to this special class of pressure-curves, which correspond to the case of expansion within an envelope impermeable to heat, the name of adiabatic curves (from $\delta\iota\alpha\beta\alpha\iota\nu\epsilon\iota\nu$, to pass through). (Clausius, 1875, p. 68)

The k in the Clausius' equation above is equal to the ratio of the heat capacities. As shown in standard texts, such as Iribarne and Godson or Bohren and Albrecht, three different statements are available. They may be derived from

Equations (5.26) and (5.27) for an adiabatic process ($Q = 0$):

$$Tv^{\frac{R}{c_v}} = \text{constant}, \tag{5.28}$$

$$Tp^{-\frac{R}{c_p}} = \text{constant}, \tag{5.29}$$

$$pv^{\frac{c_p}{c_v}} = \text{constant}. \tag{5.30}$$

These are known as *Poisson's equations*. Siméon-Denis Poisson (1781–1840) from Loiret was admitted to the École Polytechnique in 1798 and became a teacher there after graduating in 1800. Besides producing a large number of publications in mathematical physics he also played an import advocacy role in the French national educational system. Unfortunately, partly due to his failure to cite other scientists whose findings he used, he often had disagreements with his contemporaries and his work was not fully acknowledged during his lifetime.

Poisson developed the above equations in extending some of Laplace's work. Applied between two sets of conditions Equation (5.29) becomes

$$\frac{T_0}{T} = \left(\frac{p_0}{p} \right)^{\frac{R}{c_p}}.$$

By convention we choose p_0 to be 100 kPa then T_0 defines what is known as *potential temperature* which is usually denoted by θ,

$$\theta = T \left(\frac{100\,\text{kPa}}{p} \right)^{\frac{R}{c_p}}. \tag{5.31}$$

If the unsaturated atmosphere is assumed to be an ideal gas, with $R_d/c_p = 0.286$, Equation (5.31) defines the temperature that a parcel would attain were it brought to 100 kPa under adiabatic conditions. Often this is presented in the form of a rate of change of temperature with height, known as the *dry adiabatic lapse rate*, a constant,

$$\left(\frac{dT}{dz} \right)_d \equiv \Gamma_d = -\frac{g}{c_p} = -0.00975\,\text{K m}^{-1} \approx 10°\text{C km}^{-1}. \tag{5.32}$$

Thus, for every kilometer of adiabatic ascent of a non-saturated parcel the temperature declines approximately ten degrees celsius and the potential temperature remains the same.

Our thermodynamic diagram (pvT) now displays a fourth variable, θ, adiabats, with the different isopleths called isobars, isochores, isotherms, and adiabats, each having a magnitude of pressure, volume, temperature, and potential temperature respectively. These are drawn on two-dimensional and a three-dimensional indicator diagrams in figure 5.3. The steps in the Carnot cycle

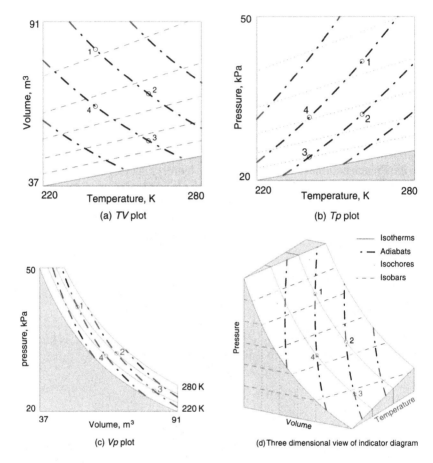

Figure 5.3 Indicator diagrams. In each view the symbols remain the same. Temperature ranges from 220 K to 280 K and pressure ranges from 20 kPa to 50 kPa. These limits produce a range of 37 m³ to 91 m³ for volume and 268 K to approximately 416 K for potential temperature. The characteristics of the real atmosphere between 20 kPa and 50 kPa fall well within these ranges

are also marked. Note that the Carnot cycle is three-dimensional on the pvT diagram.

5.8   The Second Law of Thermodynamics

As mentioned in section 5.5.2 there are an infinite set of paths for a process to move from one state to another (that may be illustrated by lines on an indicator diagram). As Carnot had shown, theoretically it is possible to control a process

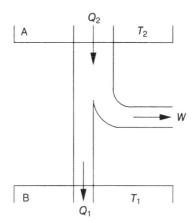

Figure 5.4 Energy flow

whereby a prescribed path follows infinitely small steps so that equilibrium is always maintained. Such a prescribed path may be retraced and therefore the process is called *reversible*. Natural systems generally are not reversible. The temperature of a hot object goes down as the temperature of surrounding objects increase slightly according to the First Law. The reverse does not occur even though energy would still be conserved.

Carnot (1824) was the first to state that in the production of mechanical work, heat (then thought of as a kind of matter) passed from a hotter into a colder body and that work was consumed in the reverse process (Clausius, 1875). Using Clapeyron's diagram [figure 5.2(b)] Clausius elaborated on the energy flows which are shown in a different form in figure 5.4. Carnot's Cycle produced work by expansion as thermal energy flowed, Q_2, from reservoir A at temperature T_2 in figure 5.4 (proportional to area 12vu in figure 5.2(b)). Additional work was performed by expansion (proportional to area 23wv) but no thermal energy flowed (adiabatic). Then work was consumed by compression through thermal energy flow, Q_1, into reservoir B at temperature T_1 (proportional to area 34wx). Additional compression consumed work (proportional to area 41xu) (again no thermal flow). Overall the work produced, W, was equal to the difference between the energy flows $Q_2 - Q_1$. The question Clausius (1875) asked was:

Whether the quantity of heat converted into work, or generated out of work, stands in a generally constant proportion to the quantity which passes over from the hotter to the colder body, or vice versa; or whether the proportion existing between them varies according to the nature of the variable body, which is the medium of the transfer.

Applying the relationship between the variables in Equation (5.28) Clausius in 1854 showed that

$$\frac{Q_1}{T_1} + \frac{Q_2}{T_2} = 0 \tag{5.33}$$

and that, in a reversible cyclic process, in his notation,

$$\int \frac{dQ}{T} = 0, \tag{5.34}$$

$$\frac{dQ}{T} = dS. \tag{5.35}$$

In Bohren and Albrecht's notation this changes to

$$\int \frac{Q}{T} dt = 0, \tag{5.36}$$

$$\frac{Q}{T} = \frac{dS}{dt}. \tag{5.37}$$

The quantity S was defined as *entropy* by Clausius in 1865 according to Gibbs (1872, p. 310):

> The term *entropy*, it will be observed, is here used in accordance with the original suggestion by Clausius, and not in the sense in which it has been employed by Professor Tait and others after his suggestion. The same quantity has been called by Professor Rankine the *Thermodynamic function*. See Clausius, Mechanische Wärmetheorie, Abhnd. ix, §14; or Pogg. Ann., Bd. cxxv (1865), p. 390; Rankine, *Phil. Trans.*, **144**, p. 126.

Clausius also extended his analysis to non-reversible processes and showed that

$$\int \frac{dQ}{T} \le 0. \tag{5.38}$$

From Equations (5.29) and (5.27) we can show that entropy is related to potential temperature,

$$dS = c_p \, d \ln \theta. \tag{5.39}$$

This is the form that we normally use for the Second Law in meteorology. Isopleths of potential temperature are also isopleths of entropy and were named *isentropic* lines by Gibbs (1872, p. 311):

> These lines are usually known by the name given them by Rankine, *adiabatic*. If, however, we follow the suggestion of Clausius and call the quantity *entropy*,

which Rankine called the *thermodynamic function*, it seems natural to go one step farther, and call the lines in which this quantity has a constant value *isentropic*.

Although the basis of entropy was developed in relation to the Carnot Cycle, its use and meaning has been extended throughout the physical and social sciences. In physical science it indicates the natural direction of energy flow. For example, thermal energy is dispersed; it becomes disorganized. In 1877 it was shown by Boltzmann to be proportional to the logarithm of the probability of the arrangement of a system. The entropy of the universe tends towards a maximum, that is natural presses tend to remove inequalities.

Absolute Temperature Scale

The towering figure of physics in the UK in the mid- to late-nineteenth century was William Thompson (1824–1907). Born in Belfast, the son of a professor of engineering, he was educated at home before attending the University of Glasgow, where he was much influenced by the French scientists of the first third of the century (e.g. Fourier, Fresnel, Lagrange, Coulomb, Laplace, Legendre, Poisson, etc.). Between 1841 and 1845 he attended Cambridge University and visited France and the scientists of the time (e.g. Cauchy, Sturm, Biot, Dumas, and Regnault). Thompson was professor at Glasgow from 1846 until 1899 and became Baron Kelvin of Largs. Of his many contributions to science his definition of absolute temperature is perhaps the most frequently quoted.

He used Equation (5.33) with one temperature as 273.16 to define the kelvin or thermodynamic temperature T,

$$T = 273.16 \frac{|Q|}{|Q_1|}, \tag{5.40}$$

where Q_1 is the thermal energy flow from or to a reservoir at the *triple point* of water, in a reversible cycle between this source and any other one, to which correspond the values of Q and T (Iribarne and Godson, 1973). The triple point is where vapor, liquid, and solid states exist in equilibrium. For water this is 273.16 K.

5.9 Water

5.9.1 Latent Heats

In the fusion/melting (subscript f), evaporation/vaporization (subscript vap) and sublimation (subscript s) of water, energy is taken up by the molecules and is released again during the reverse process (freezing, condensation, and

Table 5.4 Approximate latent heats, vaporization, fusion and sublimation as a function of temperature $MJ\,kg^{-1}$ which are different from table 5.5. Data converted to SI units from List (1958)

°C	l_{vap}	l_f	l_s
−50	2.637	0.204	2.840
−40	2.605	0.236	2.841
−30	2.577	0.264	2.841
−20	2.551	0.289	2.840
−10	2.527	0.312	2.839
0	2.503	0.334	2.837
5	2.491		
10	2.479		
15	2.467		
20	2.455		
25	2.444		
30	2.432		
35	2.420		
40	2.408		

sublimation). These are typically called *latent heats* but Bohren and Albrecht (1998) use *latent enthalpies*. The amounts of energy involved in the conversions are functions of temperature and are represented by l_f, l_{vap}, and $l_s = l_f + l_{vap}$ in $J\,kg^{-1}$, respectively. Their approximate magnitudes are given in SI units in table 5.4. (Note that at $0°C$ the magnitudes in tables 5.5 and 5.4 differ, with the Iribarne and Godson (1973) numbers being the most accurate.)

At the *triple point* of water, *tau*, the characteristics are given in table 5.5.
Iribarne and Godson give a table (IV-5) showing the variation of the specific heats with temperature.
For the total amount of energy involved in latent form per unit mass the amount of water must be multiplied by the latent heat,

$$LE = lq \ J\,kg^{-1}. \tag{5.41}$$

where q is the specific humidity defined below.

5.9.2 Expressions for Humidity

Vapor Pressure

The partial pressure of water vapor is typically represented by e in pascals.

Table 5.5 Characteristics of water vapor

Characteristic	Magnitude	Units
e_τ	610.7	Pa
T_τ	273.16	K
$\rho_{i,\tau}$	917	$kg\,m^{-3}$
$v_{i,\tau}$	1.091×10^3	$m^3\,kg^{-1}$
$v_{liq,\tau}$	1.000×10^3	$m^3\,kg^{-1}$
$v_{vap,\tau}$	206	$m^3\,kg^{-1}$
$c_{p,vap}$	1870	$J\,kg^{-1}\,K-1$
$c_{v,vap}$	1410	$J\,kg^{-1}\,K-1$
l_{vap}	2.5008×10^6	$J\,kg^{-1}$
l_s	2.8345×10^6	$J\,kg^{-1}$
l_f	0.3337×10^6	$J\,kg^{-1}$

Source: Iribarne and Godson (1973)

Absolute Humidity

Absolute humidity is the density of water vapor, ρ_{vap}, in $kg\,m^{-1}$.

Mixing Ratio

The mixing ratio, r, is the ratio of the mass of water vapor to the mass of dry air in which it is mixed,

$$r = \frac{M_{vap}}{M_d} = \frac{\rho_{vap}}{\rho_d} \; g\,kg^{-1}. \tag{5.42}$$

Specific Humidity

The specific humidity, q, is the ratio of the water vapor to the total mass of air

$$q = \frac{M_{vap}}{M_d + M_{vap}} = \frac{\rho_{vap}}{\rho_d + \rho_{vap}} = \frac{r}{1+r} \; g\,kg^{-1}. \tag{5.43}$$

Saturation

Saturation, when applied to water vapor, refers to the equilibrium situation where the same number of water molecules are leaving as are returning to a plane liquid water or ice surface. Generally this is a theoretical situation in the atmosphere because there are few plane surfaces: most are curved. Nevertheless, such a reference is useful.

Below the freezing point the saturation vapor pressure is different depending upon whether the plane surface is liquid or ice: the equilibrium is at a lower pressure over ice. Bergeron (1935) discovered in about 1919 that this leads to rapid condensation when ice is introduced into a liquid cloud, now called the Bergeron or three-phase process.

The presence of the other atmospheric gases have no influence upon the equilibrium situation. Therefore, statements about air having a capacity to hold water are quite misleading. The saturation vapor pressure, whether over liquid or over ice, depends only upon temperature.

Carnot, Clapeyron, and Clausius each investigated this dependence of the saturation vapor pressure on temperature. The equation that relates the two is

$$\frac{de_s}{dT} = \frac{1}{T}\frac{l_{vap}}{(v_{vap} - v_{liq})}. \tag{5.44}$$

A similar equation may be written for fusion. It is known as the *Clausius–Clapeyron Equation*. In order to integrate this equation and obtain the vapor pressure as a function of temperature the latent heat must be known as function of temperature. In 1946, Goff and Gratch (1946) used the Clausius–Clapeyron Equation and empirical data to produce the following formulae, which were adopted the following year by the predecessor of the World Meteorological Organization. Over plane water their equation is

$$\log_{10} e_w = -7.90298(T_s/T - 1) + 5.02808 \log_{10}(T_s/T)$$
$$- 1.3816 \times 10^{-7}(10^{11.344(1-T/T_s)} - 1)$$
$$+ 8.1328 \times 10^{-3}(10^{-3.49149(T_s/T-1)} - 1) + \log_{10} e_{ws}, \tag{5.45}$$

and, over plane ice it is

$$\log_{10} e_i = -9.09718(T_0/T - 1) - 3.56654 \log_{10}(T_0/T)$$
$$+ 0.876793(1 - T/T_0) + \log_{10} e_{i0}, \tag{5.46}$$

where

e_w = saturation vapor pressure over a plane surface of pure ordinary liquid water mb;

e_i = saturation vapor pressure of a plane surface of pure ordinary water ice mb;

T = absolute (thermodynamic) temperature K;

T_s = steam-point temperature (373.16 K);

T_0 = Ice-point temperature (273.16 K);

e_{ws} = saturation pressure of pure ordinary liquid water at steam-point temperature (1 standard atmosphere = 1013.246 mb) and

e_{i0} = saturation pressure of pure ordinary water ice at ice-point temperature (0.0060273 standard atmosphere = 6.1071 mb.)

Bohren and Albrecht (1998) give the following simpler relationships:

$$\ln \frac{e_s}{e_{s0}} \approx 19.83 - \frac{5417}{T}, \tag{5.47}$$

and

$$\ln \frac{e_s i}{e_{s0}} \approx 22.49 - \frac{6142}{T}, \tag{5.48}$$

where $e_{s0} = 611$ Pa, the saturation vapor pressure at 273.16 K. Saturation vapor pressures are not dependent upon total pressure.

The terms *evaporation* and *condensation* are used to describe situations where the net flow of molecules relative to the liquid produces a loss or gain.

Formulae and tables may be derived for the saturation mixing ratios, $r_{s,vap}$ and $r_{s,i}$, and for saturation specific humidities, $q_{s,vap}$ and $q_{s,i}$.

At sea level at a temperature of $0°C$ the saturation vapor pressure of about 6 mb coincides with a saturation mixing ratio and a specific humidity of about $4\,g\,kg^{-1}$.

Relative Humidity

Relative humidity, RH, having a magnitude between 0.0 and 1.0, is given by

$$RH = \frac{e}{e_s} \approx \frac{r}{r_s} \approx \frac{q}{q_s}. \tag{5.49}$$

The relative humidity over ice is calculated with e_{si}. In order to obtain relative humidity in percentages Equation (5.49) must be multiplied by 100.

5.9.3 Equation of State for Moist Air

Water has a molecular weight of $m_{vap} = 18.015$. Therefore its equation of state is

$$e = \rho_{vap} \frac{R^*}{m_{vap}} T = \rho_{vap} R_{vap} T. \tag{5.50}$$

The equation of state for moist air may be obtained by the addition of the two separate pressures p_d and e following Dalton's Law [Equation (5.4)] and manipulation of the molecular weights,

$$p = p_d + e = \rho R_d (1 + 0.608q) T = \rho R_d T_{virtual}. \tag{5.51}$$

The virtual temperature, $T_{virtual}$, is a meteorological invention so that the equation of state for moist air remains in its conventional form with the gas

Table 5.6 Virtual temperature increment

°C	100 kPa	50 kPa
−20	0.12	0.24
0	0.64	1.27
20	2.62	5.29
40	9.03	

Source: (List, 1958)

constant for dry air. The difference between the virtual temperature and temperature, $\Delta T_{virtual} = T_{virtual} - T$, is known as the *virtual temperature increment*. Table 5.6 lists the virtual temperature increment for saturated air.

Substitution of the separate equations of state into Equation (5.42) also provides a relationship between vapor pressure and mixing ratio

$$r = \frac{0.622e}{p - e}. \tag{5.52}$$

It follows that

$$q = \frac{0.622e}{p - 0.378e}. \tag{5.53}$$

5.9.4 Water and Adiabats

Estimates for the typical atmosphere reveal that ignoring water vapor in the calculation of potential temperature leads to errors of less than 0.1°C. Therefore the use of dry adiabats for moist unsaturated air is generally acceptable.

If the temperature of air declines the equilibrium vapor pressure of water also declines. When the equilibrium level is reached, changes in phase of the water will release thermal energy to the air. Consequently this process leads to new adiabats that take into account an increase in temperature offsetting the decline due to expansion. Because the liquid and/or ice will also affect the temperature calculations usually assume that all the liquid and ice fall out immediately. The reverse process in the free atmosphere is limited because liquid and/or ice would have to be present (as in clouds and beneath precipitating clouds). The new adiabats are called both *saturation adiabats* and *pseudo-adiabats*.

Following the tradition started by Watt these processes are typically explained with indicator diagrams. In meteorology several different ones have been developed specifically for the atmosphere. In particular the skew-T is used often in the US while the tephigram is favored by the UK and Commonwealth countries. Today these can be displayed by computer but it is the equations are needed in the numerical models.

5.10 Discussion

Following the traditional division of subject matter, in this chapter we have considered the variables used to describe a gas and their relationship to energy. These are, however, integral parts of atmospheric processes and should not be isolated as some independent elements. It is through a knowledge of thermodynamics that we can understand how energy flows into, through, and out of the atmosphere.

Applied to the example of the conservation of energy in section 4.3.5 we can use thermodynamics to explain how energy for work in lifting an object comes from food (chemical energy). Also, we can see how it is finally dissipated into thermal energy when it falls, i.e. some is converted through friction with the air into internal energy, but most into internal energy on impact through vibrations. The sound waves may be heard and the increase in temperature may be sensed. It is the conversion of mechanical to thermal energy. The temperature increase due to a single impact is small but a few impacts, as a hammer on a nail, can produce a hot enough nail to burn the finger (see Boyle's statement in section 5.4).

Different gases play different roles especially in radiation (see next chapter) but water is clearly unique in its importance for the whole atmospheric system.

Chapter 6

Radiation

6.1 Early Work

6.1.1 Light

In 1905 Albert Einstein asserted that light consisted of individual quanta that behaved both like particles and like waves thus melding what had until that time been two separate concepts. The particle theory originated with Pythagoras while the wave theory dates to 1665 with independent statements by Francesco Grimaldi and Robert Hooke. In that same year Newton had experimented with a prism to separate light into the colors of the rainbow. Interference, diffraction, refraction, and polarization were all studied by various authors in the mid and late seventeenth century.

Descartes discussed the variation of the speed of light in air and water in 1637 but one of the first estimates was made by the Danish astronomer, O. Römer (1644–1710). Using the moons of Jupiter he concluded that it took light 11 minutes to pass the distance of the diameter of the earth's orbit (Parkinson, 1981). In 1678 Huygens introduced the concept of a substance *ether* necessary for the transmission of light by analogy to sound waves in air. This idea was eventually put into print in his *Treatise on light* in 1690 (Huygens, 1690). Subsequently the "ether" was universally adopted as a real entity and it was not seriously challenged until 1846 when Michael Faraday (1791–1867) suggested that it was unnecessary.

Most of Newton's 40 years of optical work did not appear until 1704. He emphasized the particle approach but also recognized the presence of vibrations. The wave theory was revived by Thomas Young in the early nineteenth century and it was given strong support by Fresnel (1818) with his publication of *Memoir on the diffraction of light*.

One of the first to investigate the spectrum of substances was Thomas Melvill (1726–53). Born in Scotland he went to the University of Glasgow in 1748 to study divinity. While there he collaborated with Alexander Wilson to measure temperature as a function of altitude using kites. He studied Newton's *Opticks* and in 1752 reported on experiments in which he introduced different materials into an alcohol flame and noted their colors. The latter he thought were due to the different speeds of the corpuscles. His work apparently had little impact on future developments of the science of radiation.

That light was a combination of wavelengths was clear. In 1802 William Hyde Woolaston (1766–1828) a product of a wealthy and well-known family from East Dereham, England, and a Cambridge graduate, reported seeing dark lines in the solar spectrum. These were carefully measured in 1814 by the German scientist Joseph Fraunhofer (1787–1826), using an exceptionally good prism. It was found that other stars also had lines but at different wavelengths. In 1859 Kirchhoff suggested that these were absorption lines due to the star's outer gases and could therefore be used to identify the composition of those gases. In 1868 Anders Ångström listed 1,000 such lines in the sun's spectrum.

6.1.2 Extension to Other Wavelengths

William Herschel (1738–1822) was born in Hannover, Germany, but he traveled to England with his army officer father and learned the language as a teenager. He moved to England permanently when France occupied Hannover in 1757. His early career was as a musician but by the 1770s he had turned to astronomy for which he became widely recognized. In 1800 while studying the heat in the different colors of the spectrum Hershel discovered that there was energy beyond the red in the longer wavelengths. Thus he had identified the *infrared*. The following year Johann Ritter (1776–1810), in search for other invisible wavelengths, discovered the *ultraviolet* at shorter wavelengths than the visible.

6.2 Quanta

James Clerk Maxwell, who is important in the development of the kinetic theory of gases, had a fundamental influence on radiation theory when he published a paper entitled *On a dynamical theory of the electromagnetic field* in 1865 (Maxwell, 1865). Using the concept of the ether he predicted the existence of electromagnetic waves other than light. They traveled at the speed of light and indeed light had the form of an electromagnetic wave. In 1900 Max Planck, in his investigation of the emission and absorption of electromagnetic energy, suggested that radiation was composed of discrete units of energy. Five years later Albert Einstein went much further in his paper, *On a heuristic viewpoint concerning the generation and transformation of light.*

> On the assumption here to be considered, energy during the propagation of a ray of light is not continuously distributed over steadily increasing space, but it consists of a finite number of energy quanta localized at points in space, moving without dividing and capable of being absorbed or generated only as entities. (Einstein, 1905)

The basic idea underlying the quantum theory of radiation is that an atom can have only discrete energy levels, $E_1, E_2, \ldots, E_n, \ldots$. Each time an atom gives off energy (radiates) the energy level drops from one level to another. The difference in energy is proportional to the frequency of the radiation, that is

$$E_j - E_i \propto \nu.$$

The proportionality constant is h, *Planck's constant* ($6.6255916 \times 10^{-34}$ J s). Thus

$$E_j - E_i = h\nu,$$

or,

$$\nu = \frac{E_j - E_i}{h}. \tag{6.1}$$

Since electromagnetic radiation travels (in a vacuum) at the speed of light, c (2.997925×10^8 m s^{-1}), the relationship between wavelength and frequency is given by

$$\lambda = c\,\nu. \tag{6.2}$$

Like emission, absorption may only occur at specific frequencies peculiar to the type of atom. Gases in the atmosphere display mainly bands of absorption and emission. For a monotonic gas the quantized energy is electronic. Therefore the absorption and the emission spectrum of a monotonic gas will consist of a set of discrete lines at particular frequencies (or wavelengths). For polyatomic gases additional quantized energy is in the form of vibrational and rotational energies. The combined effect of the different quantized energies for a substance is to produce groupings of the individual lines into a band spectrum. Also, if the gas contains many molecules then their interaction will affect the frequencies slightly, thus

Table 6.1 The electromagnetic spectrum

Name	Typical wavelength m	Typical frequency $s^{-1} = Hz$	Main source of energy
X-rays	10^{-9}	3×10^{17}	Electronic
Ultraviolet	10^{-7}	3×10^{15}	Electronic
	0.38×10^{-6}		
Visible	0.5×10^{-6}	6×10^{14}	Vibration
	0.78×10^{-6}		
Infrared	10^{-6}	3×10^{14}	Vibration
Microwaves	10^{-3}	3×10^{11}	Rotation
TV	1	3×10^{8}	Magnetism
AM Radio	10^{2}	3×10^{6}	Magnetism

broadening a line. This is known as *pressure broadening*. In a solid or liquid, where there is strong interaction between neighboring molecules, the absorption and emission spectra are continuous and are described by Planck's Equation (6.21).

Some of the current terms used in the electromagnetic spectrum along with their typical wavelengths and frequencies are given in table 6.1. Also included are the main sources of energy for these divisions. For atmospheric studies solar radiation is typically considered to include parts of the ultaviolet and infrared as well as the visible. Most earth-bound objects naturally radiate in the infrared.

Translational energy is not quantized. Consequently radiation does not affect temperature directly. However, because all forms of energy in a gas at thermodynamic equilibrium are partitioned according to a definite scheme, radiation does lead to temperature change. For example, a diatomic gas, visualized as a simple dumbbell, will have rotational and vibrational energies as well as translational energy. On collision with another molecule the translational energy of the second molecule will be modified by all three energies of the first, and vice versa.

It should be noted that the term, *radiation* is used to refer both to the form of the energy and to the process. The adjective *monochromatic* is applied to radiation of a specific wavelength and variables that apply to such radiation are hereafter subscripted by λ. When radiation is given in terms of frequency the subscript v is used. Sometimes wave number, n in units of m^{-1}, the inverse of wavelength, the number of waves per unit length, is used instead of wavelength or frequency.

6.3 Definitions and Laws of Radiation

6.3.1 Bouger's Law

Pierre Bouger (1698–1758) at the age of 15 succeeded his father as royal professor of hydrography in France. He was an authority on all things nautical but more importantly for our consideration of radiation he showed, in 1729 in his *Optical essay on the gradation of light* (Bouger, 1729), that there was an exponential decrease of light in a beam passing through an absorbing medium of uniform transparency.

For a single absorbing medium, the absorption fraction, a_λ, is proportional to the density, ρ, of the absorbing material and the distance within the material, dl,

$$a_\lambda = \frac{dI_\lambda}{I_\lambda} = -k_\lambda \, \rho \, dl. \tag{6.3}$$

Integration of Equation (6.3) gives

$$I_\lambda = I_{\lambda 0} e^{-\int_0^l k_\lambda \, \rho \, dl} \ \mathrm{W \, m^{-2} \, m^{-1} \, sr^{-1}}. \tag{6.4}$$

If the material is homogeneous, i.e. k_λ is independent of l,

$$I_\lambda = I_{\lambda 0} e^{-k_\lambda \int_0^l \rho \, dl}. \tag{6.5}$$

The integral, $\int_0^l \rho \, dl$, which is the amount of material in a column of length l and unit cross-section, is often represented by "u" and is called the *optical depth* or *optical thickness*,

$$I_\lambda = I_{\lambda 0} e^{-k_\lambda u}. \tag{6.6}$$

This is *Bouger's Law*. Probably because Bouger's *Essai* was not widely available, it was restated in 1760 by Lambert, who often gets credit for it. Also, Equation (6.6) is sometimes known as *Beer's Law* after a German banker and amateur but, accomplished, selenographer of the early nineteenth century.

6.3.2 Lambert's Law

Johann Heinreich Lambert (1728–77), the son of a poor Alsace tailor, gave up school at twelve in order to work for his father. However, he continued his studies at night and, because of his excellent handwriting, found a job as a secretary to a scholar who became a professor of Law at Basel University. He later became a children's tutor (1748–58) to a wealthy family which had accumulated a large library where Lambert studied in his spare time. He became very knowledgeable in mathematics, physics, and philosophy. For his own interest he recorded both astronomical and meteorological observations.

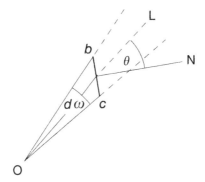

Figure 6.1 Lambert's cosine law. The geometry of radiation for a small planar area, dA, which is shown edge on as bc and is contained in a small solid angle $d\omega$, where N is the normal to the plane. The area perpendicular to the line of sight OL is given by $dA \cos\theta$

In 1760 he published *Photometry, or On the measurement and grading of light, color, and shade* (Lambert, 1760). In it he described his discovery that the brightness of a diffusely radiating plate is proportional to the cosine of the angle between the line of sight and the normal to the surface. It is not affected by the orientation around the normal (N). Of course, this trigonometric function also describes the area that such a unit plate presents perpendicular to the line of sight (figure 6.1). This is known as *Lambert's Cosine Law*. Radiation that obeys this law is therefore given the adjective *Lambertian*. The law is exact only for an ideal object, known as a *black body* defined by Kirchhoff and Planck (see below).

Lambert did not go unnoticed in the scientific community and in 1764 he was offered a position in Berlin where he was welcomed by a number of academicians including Euler. Apparently his appearance and behavior were unusual so Frederick the Great did not give him an appointment for a year (Scriba, 1981).

6.3.3 Intensity

The *intensity* of radiation is defined as the amount of radiant energy, E, passing through a unit area per unit time from, or to, a unit angle of direction. From figure 6.1 this may be written

$$I_{\theta,\psi} = \frac{d^3 E}{dA \cos\theta \, dt \, d\omega} \tag{6.7}$$

in $\text{W m}^{-2}\,\text{sr}^{-1}$ where the solid angle, $d\omega$, is given in terms of $d\theta$ and $d\psi$.

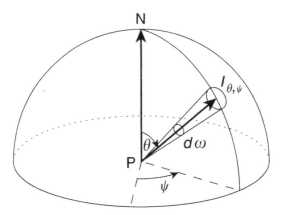

Figure 6.2 Geometry of radiative intensity, $I_{\theta,\psi}$, for a small planar area, dA, at P, where N is the normal to the plane

In figure 6.2 the intensity is shown as a vector, $I_{\theta,\psi}$. This vector could also point in the opposite direction. It passes through a unit area at P, dA (not shown), whose normal vector is N. The subscripts on I refer to the angle from the normal (θ), and the angle from some reference direction in the dA plane (ψ). The I vector represents a cone of solid angle $d\omega$. Some authors display the area dA so the apex of the cone is projected beyond P (as in figure 6.1). Then they may call the volume a "pencil."

6.3.4 Flux

The *radiant flux density* or *irradiance*, F, is the flow of radiant energy through a unit area in a unit time, in units of W m^{-2}. It is the integration of intensity over all directions. From Equation (6.7)

$$F = \int I_{\theta,\psi} \cos\theta \, d\omega = \frac{d^2 E}{dA \, dt} \qquad (6.8)$$

in W m^{-2}. For the integration, the solid angle $d\omega$ may be represented using spherical coordinates (see figure 6.3). The area of the element at A is

$$d\alpha = r \, d\theta \times r \sin\theta \, d\psi = r^2 \sin\theta \, d\theta \, d\psi.$$

If $d\alpha$ is substituted into the differential form of Equation (2.6) we obtain

$$d\omega = \sin\theta \, d\theta \, d\psi. \qquad (6.9)$$

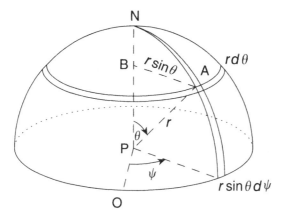

Figure 6.3 Solid angle in spherical coordinates. The element of area at A is $d\alpha = r\,d\theta \times r\,\sin\theta\,d\psi$. Inserted into the differential form of Equation (2.6) the latter becomes $d\omega = \sin\theta\,d\theta\,d\psi$

Therefore, the flux density for one hemisphere may be written

$$F = \int_0^{2\pi} \int_0^{\pi/2} I_{\theta,\psi} \cos\theta \sin\theta \, d\theta \, d\psi \qquad (6.10)$$

in $W\,m^{-2}$.

6.3.5 Special Cases

In the special case where the radiant intensity is independent of direction, called *isotropic* intensity, Equation (6.8) reduces to

$$F_i = \pi I. \qquad (6.11)$$

So far, both the intensity and the irradiance have been considered to be independent of wavelength. However, the same equations may be rewritten to express the *monochromatic* intensity, $I_\lambda = dI/d\lambda$, and monochromatic flux density (or monochromatic irradiance or spectral flux density), $F_\lambda = dF/d\lambda$. The units of monochromatic irradiance are $W\,m^{-2}\,m^{-1}$.

The relationships for specific frequencies are given by

$$I_\lambda = -\frac{c}{\lambda^2} I_\nu, \qquad F_\lambda = -\frac{c}{\lambda^2} F_\nu, \qquad (6.12)$$

where the units are $W\,m^{-2}\,m^{-1}\,sr^{-1}$ and $W\,m^{-2}\,m^{-1}$ respectively. The negative signs indicate that as wavelength increases frequency decreases.

For parallel radiation the above relationships between intensity and flux do not hold.

6.3.6 Kirchhoff's Law

Rosenfeld (1981) has argued that economic circumstances pushed the Germans to the limits of scientific knowledge in the nineteenth century. As a result they were the ones who made discoveries in radiation during this era.

Late-comers in the industrial revolution, the Germans, more than their wealthier English and French competitors, had to rely on scientific methods for the improvement of technology. The paradoxical result was that physics and chemistry found more favorable conditions of development under the multitude of feudal governments in Germany than in the progressive environments of prosperous manufacturing centers in England and France.

Gustav Robert Kirchhoff (1824–87), the most gifted son of a law councillor in Königsberg, was influenced at the local university by Franz Neumann and his ideas concerning electromagnetism. He became a professor at Breslau in 1851. There he met and became friends with Bunsen who moved to Heidelberg. In 1854, at Bunsen's instigation, Kirchhoff accepted a professorship of physics at Heidelberg where he stayed for 21 years. In 1875 he moved to the chair of theoretical physics in Berlin.

Despite an accident which forced him to use crutches and a wheelchair and a long illness later in life, Kirchhoff had an outstanding scientific career. Due to the influence of Neumann his early work was in electromagnetism. This led to an interest in optical phenomena which seemed to have analogous characteristics. While Bunsen was analyzing salts using a flame and colored filters Kirchhoff suggested the application of spectral analysis. By this method in 1860 Bunsen and Kirchhoff were able to establish a new and precise means of identifying substances by their unique line spectra.

During that research Kirchhoff noticed lines in the sodium spectrum which coincided with the solar dark D lines of Fraunhofer. He interpreted these dark lines as being due to the absorption by sodium in the Sun's upper atmosphere. The theory permitted the recognition of chemical components of the sun and other stars. In 1859 he generalized his findings, stating that: for rays of the same wavelength at the same temperature the ratio of the emissive power to the absorptive power is the same for all bodies (Kirchhoff, 1859),

$$\frac{I_{\lambda T}}{A_{\lambda T}} = c_{\lambda T}. \qquad (6.13)$$

where $c_{\lambda T}$ is a constant. This is one form of *Kirchhoff's Law*. Three years later he introduced the concept of a *black body* which absorbs all the radiation incident upon it. The application of Equation (6.13) to such an object leads to the theoretical emission spectrum that was finally defined precisely by Kirchhoff's successor to the Berlin chair, Max Planck (see below).

Another way of writing Equation (6.13) using this theoretical black body emission, $I_{\lambda T}^*$, (* designates a black body) is

$$\frac{I_{\lambda T}}{a_{\lambda T}} = I_{\lambda T}^* \text{ W m}^{-2} \text{ m}^{-1} \text{ sr}^{-1}, \tag{6.14}$$

where $a_{\lambda T}$ is the fractional absorption.

Also, if $\epsilon_{\lambda T}$ is the fractional emission

$$I_{\lambda T} = \epsilon_{\lambda T} I_{\lambda T}^*,$$

and

$$\frac{\epsilon_{\lambda T} I_{\lambda T}^*}{a_{\lambda T}} = I_{\lambda T}^* \text{ W m}^{-2} \text{ m}^{-1} \text{ sr}^{-1},$$

then

$$\epsilon_{\lambda T} = a_{\lambda T}. \tag{6.15}$$

which is the usual form of *Kirchhoff's Law*. For a black body

$$\epsilon_{\lambda T} = a_{\lambda T} = 1. \tag{6.16}$$

Since ϵ_λ is usually constant over a wide range of wavelengths for a solid or a liquid body, a representative overall fractional emission may be defined by

$$\epsilon = \frac{\int_0^\infty \epsilon_\lambda F_\lambda \, d\lambda}{F^*}, \tag{6.17}$$

where F^* is the irradiance from a black body.

An object that displays fractional absorption and emission is called a *gray body*.

6.3.7   Stefan–Boltzmann's Law

The Austrian Josef Stefan (1835–93), whose parents came from Slovakia, was an excellent teacher and experimenter. In 1863 he became professor of higher mathematics and physics at the University of Vienna. He was also appointed director of the Institute for Experimental Physics, a unit that Doppler had established in 1850. There Stefan worked on light, sound, and electricity. In reviewing the experimental work of the French chemists Dulong and Petit, who had found that the total quantity of heat radiated in a unit time by a body at temperature T was proportional to $(1.0077)^T$, Stefan instead proposed that it was proportional to the fourth power of absolute temperature of the body, the black body irradiance, F^*, (Stefan, 1879),

$$F^* \propto T^4. \tag{6.18}$$

Four years later in 1883 Ludwig Boltzmann worked "out a theoretical derivation, based on the second law of thermodynamics and Maxwell's electromagnetic theory, of the fourth power law ... of Stefan" (Brush, 1981). Thus the full equation is given by

$$F^* = \sigma T^4 \, \text{W m}^{-2}, \tag{6.19}$$

where $\sigma = 5.66961 \times 10^{-8} \, \text{W m}^{-2} \, \text{K}^{-4}$. Equation (6.19) is known as *Stefan–Boltzmann's Law*.

Boltzmann (1844–1906), also an Austrian having been born in Vienna, studied with Stefan. As a professor at Graz, Vienna, Munich, and Leipzig, he worked for 40 years on the kinetic theory of Clausius and Maxwell. That field, however, accounted for only about half his publications. The rest were on a broad range of topics in physics, chemistry, mathematics, and philosophy (Brush, 1981). He related entropy to probability (1877) and introduced the concept of ergodicity (see section 3.4.1).

6.3.8 Wien's Law

Wilhelm Carl Werner Otto Fritz Franz Wien (1864–1928), born as the only child of an East Prussian farming family, had little academic success early in life. He gave up his first attempt at university in Göttingen but found farming also not to his liking. He was an introvert and apparently did not like the many social activities of students. However, eventually he entered the University of Berlin in 1883 where he was excited by von Helmholtz. Amazingly he received a doctorate only three years later with a dissertation on diffraction by a grating. He then returned to his family's farm where he was free to study theoretical physics independently. In 1888 von Helmholtz had become director of Physikalisch-Technische Reichsansalt (PTR) and Wien became associated with it. Between 1894 and 1897 at Helmoltz's suggestion, he worked on the theory of sea waves and cyclones (Kangro, 1981b). He married in 1898. From 1899 to 1919 he was a full professor at Würzburg. Wien traveled widely and in 1913 lectured at Columbia University. His final years were spent at the University of Munich where he was rector.

Wien worked on vacuum tubes and electromagnetism. He conceived of cavity radiation and then partially abandoned it. In 1893 he demonstrated for radiation that $\lambda \times T$ was a constant. This is often expressed today as "the wavelength of maximum intensity of radiation from a black body is inversely proportional to temperature," i.e.

$$\lambda_{max} = \frac{a_{\lambda,max}}{T},$$
(6.20)

where $a_{\lambda,max} = 2897.0 \times 10^{-6}$ m K. This is known as *Wien's displacement law* or just *Wien's Law*.

He also developed a formula for black body emission which was very close to Planck's equation.

6.3.9 Planck's Equation

It has already been mentioned that Planck followed Kirchhoff both as professor in Berlin and in deriving a mathematical expression for the emission from the latter's concept of an ideal body, a black body.

Max Karl Ernst Ludwig Planck (1858–1947) was the fourth child of a professor of civil law in Kiel, Germany. He was talented in music and mathematics and almost selected the former as a career. He studied in Berlin under Kirchhoff and von Helmholtz and independently read Clausius's works. His doctorate was obtained at the University of Munich in 1879. In Munich he met his future wife and became friends with the mathematician Carl Runge. At school he had been introduced to the concept of the conservation of energy which much influenced his thinking (Kangro, 1981a). In 1885 he was appointed professor at Kiel and succeeded Kirchhoff in Berlin in 1888. He remained there until 1926. His work was mainly in theoretical physics applied to thermodynamics, radiation, and relativity. He also wrote on the philosophy of physics.

Planck derived Wien's distribution law and, around 1900, his own equation for a black body,

$$I_\lambda^* = \frac{2hc^2}{\lambda^5} \left(\frac{1}{e^{[hc/\lambda kT]} - 1} \right) \text{W m}^{-2} \text{m}^{-1} \text{sr}^{-1},$$
(6.21)

or, in terms of frequency,

$$I_\nu^* = \frac{2h\nu^3}{c^2} \left(\frac{1}{e^{[h\nu/kT]} - 1} \right) \text{W m}^{-2} \text{s sr}^{-1},$$
(6.22)

or, in terms of wavenumber,

$$I_n^* = 2hc^2 n^3 \left(\frac{1}{e^{[hcn/kT]} - 1} \right) \text{W m}^{-2} \text{m sr}^{-1},$$
(6.23)

where the black body monochromatic intensity at wavelength λ, I_λ^*, is given in W m^{-2} m^{-1} sr^{-1},
or the black body unit frequency intensity at frequency ν,

I_v^*, is given in $\text{W m}^{-2}\,\text{s sr}^{-1}$,
or the black body unit wavenumber intensity at wavenumber n,
I_n^*, is given in $\text{W m}^{-2}\,\text{m sr}^{-1}$,
h is Planck's constant $= 6.6255916 \times 10^{-34}\,\text{J s}$,
c is the speed of light $= 2.997925 \times 10^8\,\text{m s}^{-1}$,
k is Boltzmann's constant $= 1.380546 \times 10^{-23}\,\text{J K}^{-1}$,
T is temperature in kelvin $= \text{K}$,
λ is wavelength in meters $= \text{m}$,
v is frequency in per second $= \text{s}^{-1}$, and
n is is the wavenumber in per meter $= \text{m}^{-1}$.
Sometimes a negative sign appears in Equations (6.22) and (6.23) to show that
the axes of v and n are reversed from λ. The constants may be combined, for
example in Equation (6.21):

$$2hc^2 = c_1 = 1.191066 \times 10^{-16}\,\text{J m}^2\,\text{s}^{-1},$$

$$hc/k = c_2 = 1.438833 \times 10^{-2}\,\text{m K}.$$

Thus,

$$I_\lambda^* = \frac{c_1}{\lambda^5} \left(\frac{1}{e^{[c_2/\lambda T]} - 1} \right)\ \text{W m}^{-2}\,\text{m}^{-1}\,\text{sr}^{-1}. \tag{6.24}$$

Two examples of Planck's Equation at temperatures of 250 K and 325 K are
drawn in figure 6.4. Since Stefan–Boltzmann's Law is the total energy emitted from a black body, it is derivable from Planck's equation by integration
with respect to angle and to wavelength. For isotropic radiation the former is
simply a multiplication by the constant, π, as in Equation (6.11). Integration
by wavelength (an interesting exercise) is proportional to the area under the
Planck curve (for example in figure 6.4). Thus Stefan–Boltzmann's constant
is shown to be

$$\sigma = \frac{2\pi^5 k^4}{15c^2 h^3}. \tag{6.25}$$

Similarly, Wien's Law is derivable by differentiation.

It should be noted that for a specified temperature, Equations (6.21), (6.22),
and (6.23) give different distributions such that the wavelength of maximum
intensity is different. For example, Wien's Law applied to Equation (6.22),
becomes

$$\lambda_{v,max} = \frac{a_{v,max}}{T}, \tag{6.26}$$

where $a_{v,max} = 5096.8 \times 10^{-6}\,\text{m K}$ (Valley, 1965, appendix B).

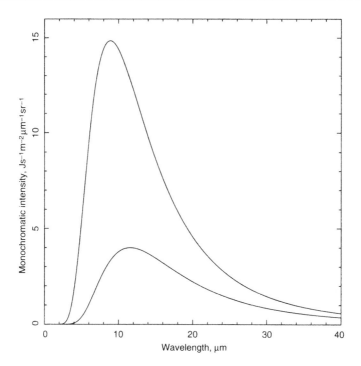

Figure 6.4 Planck's curves for 250 and 325 K. From Stefan–Boltzmann's Law the total energies emitted are 221.5 W m^{-2} and 632.5 W m^{-2}. The wavelengths of maximum intensities from Wien's Law are 11.6 μm and 8.9 μm respectively

6.3.10 Schwarzschild's Equation

Now that we have equations for emission and absorption we can consider their net result as a function of angle and wavelength. The infinitesimal optical depth dl through a slab, such as a gas, varies as a function of the angle to the normal or, in the atmosphere, as a function of the zenith angle z (see figure 6.5). With substitution Equation (6.3) becomes

$$a_\lambda = \frac{dI_\lambda}{I_\lambda} = -k_\lambda \rho \sec \theta \, dz. \qquad (6.27)$$

Then if $\rho \, dz$ is replaced from the hydrostatic equation (7.53) this becomes

$$a_\lambda = \frac{dI_\lambda}{I_\lambda} = k_\lambda \sec \theta \frac{dp}{g}. \qquad (6.28)$$

When monochromatic radiation passes through a layer of absorbing material some intensity is removed by absorption, as given by Bouger's relation in the

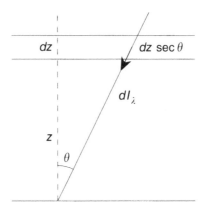

Figure 6.5 Absorption dI_λ in a layer dz thick at a zenith angle θ

form of Equation (6.28), and some will be added by emission from the same absorbing material as given by Planck's Law (Equation 6.21). The change in intensity is therefore,

$$dI_\lambda = -(I_\lambda + I^*)k_\lambda \sec\theta \, \frac{dp}{g}. \tag{6.29}$$

This is known as *Schwarzschild's Equation* after the German astronomer Karl Schwarzschild (1873–1916).

6.3.11 Scattering

The *scattering* of radiation is a process whereby particles diffuse incident radiation in multiple directions. The wavelength of the radiation remains unchanged but the wavelengths that are scattered are selective and depend mainly upon the size of the particles. Two types of scattering are usually considered relevant in the atmosphere.

In 1871 Rayleigh showed that visible wavelengths are scattered by atmospheric molecules, which have a radius less than one tenth the wavelength of the visible, as a function of the inverse of the fourth power of the wavelength of the incident radiation. His equation is

$$k_s = N\pi r^2 \frac{128\pi^4 r^4}{3\lambda^4} \left(\frac{n^2 - 1}{n^2 + 2}\right)^2, \tag{6.30}$$

where k_s is the scattering coefficient, N is the concentration of the particles, r is their radius, λ is the wavelength, and n is the index of refraction of the particles. The first term, $N\pi r^2$ is the cross-sectional area of the scatterers in a unit volume. The energy is sent equally forwards and backwards and to a lesser degree sideways along the line of receipt. These findings mean that scattering at the blue end of the visible spectrum is favored and accounts for the blue sky.

Lord Rayleigh (1842–1919) was born John William Strutt in Essex, England, and inherited his peerage of "Third Baron Rayleigh." As a result he was in a financial position to support his own life-long research into experimental and theoretical physics after he graduated from Cambridge in 1865.

Somewhat later, in 1908, Gustav Mie (1868–1957), who had received a doctorate from Heidelberg in 1891, developed a more general theory of scattering which encompassed Rayleigh scattering. With respect to larger particles, those with radii greater than one fourth of the wavelength of light, he showed that the scattering was mainly forward from the particle and was proportional to λ^{-1} (Mie, 1908).

As in the case of absorption the amount of scattering depends upon the optical depth of the scattering material. Therefore, when the sun is near the horizon and the path length is long, most of the blue part of the visible is scattered out and the dominant red remains.

Together the effects of absorption and scattering in reducing energy in a beam is called *attenuation*.

6.4 Application to the Earth

6.4.1 Solar Receipt

That solar radiation varies has long been recognized. Solar radiation at the earth varies as a result of solar output, of orbital characteristics, and of changes in those characteristics. Fascination with the heavens is evident in the life of early humans (Hawkins, 1965) and measurements of the positions of stars date from the Babylonian period (3rd millenium BC). Greek astronomers, in particular Hipparchus (ca. 200–100 BC), were the first to estimate the distances of the moon and sun and to recognize orbital changes. Based upon the laws of Kepler and Newton, Laplace and Lagrange were able to develop more precise equations detailing orbital characteristics of the earth and planets corresponding to the observed motions. A recent short treatment of celestial mechanics is given by Collins (1989).

In the 1920s the mathematician Mulitin Milankovitch (1879–1958) attempted to combine such variations into a causal explanation of climatic change. His derivations are fully described in Milankovitch (1969). Born in Dalj, Croatia, and educated at the Institute of Technology in Vienna (PhD 1904), Milankovitch was a professor at the University of Belgrade. The theory, which bears his name, was revived in the 1970s. Vernekar (1972) made more refined calculations and (Imbrie and Imbrie, 1980) popularized the theory using the latest estimates of orbital characteristics and their temporal variations obtained by Berger (1977). Recently variations in solar receipt have been incorporated into a general circulation model (Lean and Rind, 1998).

Solar Output

The energy of the sun derives from the thermonuclear reaction of hydrogen being converted into helium. The central temperature is estimated to be in the order of 1.5×10^7 K, decreasing to about 4,300 K just below the photosphere, the apparent surface of the sun, at a radius of 696,000 km. Above the photosphere is a hot, 10 to 60×10^3 K, 10×10^3 km thick transparent glowing gas known as the *chromosphere*. Below the photosphere is a deep convective layer. The opaqueness of the photosphere is due to negative hydrogen ions immediately below the photosphere. Most of the energy reaching the earth is emitted from the photosphere. Absorption by the upper photosphere produces the lines named after Fraunhofer. The gases of the sun are not in solid rotation around its axis which is tilted at $7°$ to the plane of the earth's orbit. The rotation time is 26 days at its equator and 34 days at the pole giving an effective period of 27 days (Evans, 1965). Solar radiation measured above the earth's atmosphere deviates from a black body curve (see figure 6.6). The peak intensity appears at approximately 0.475 μm. If this be substituted into Wien's Law a temperature of 6,090 K results. This is known as the *color temperature* of the sun. The average irradiance at the top of the atmosphere is estimated to be 1,376 W m^{-2} and is called the *solar constant*. The actual irradiance is quite variable with time. For example, Willson (1993) lists estimates for the twentieth century of between 1,346 and 1,389 W m^{-2}. He also gives results of satellite measurements between 1976 and 1985 that vary between 1,367.1 and 1,389 W m^{-2}. If 1,376 W m^{-2} be adjusted for the average earth's orbital distance (149.68 $\times 10^6$ km) and substituted into Stefan–Boltzmann's Law a temperature of 5,880 K is obtained. This is called the *effective temperature* of the sun.

Variation in solar output may be correlated with features that appear in the photosphere such as dark areas called *sunspots*. A record extending back to 1700 indicates an approximate 11 year cycle in such phenomena (Evans, 1965). Numerous researchers have investigated the possible relationship between sunspots and weather and climate and while statistical correlations have been claimed for specific variables in specific locations for specific periods none withstand more detailed analyses: a physical mechanism yet has to be discovered. Even without an adequate theory some researchers have gone even further and related solar cycles to social and economic series (Currie, 1988).

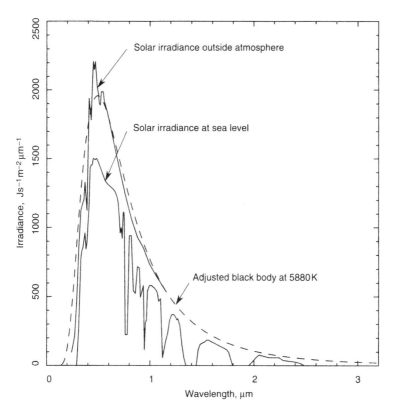

Figure 6.6 Solar irradiance outside the atmosphere and at sea level. Also included is the black body irradiance for a temperature of 5880 K adjusted for the average distance sun–earth. The gaps in the sea level irradiance are due to the absorption bands of atmospheric gases. Data modified from Gast (1965)

Eccentricity

The distance of the earth from the sun clearly affects the total irradiance at the top of the atmosphere through the year. The *eccentricity* (see section 2.1.2) is an inverse measure of the extent to which the orbit varies from a circle. An estimate of the eccentricity for 1950 is $e = 0.0167$ (Berger, 1977). At perihelion, on January 3, the distance is approximately 147.2×10^6 km and at aphelion, on July 4, the distance is 152.2×10^6 km. At the scale of figure 2.2 this ellipse would appear as a circle. The 3 percent difference in distance causes a 6 percent change in solar irradiance between aphelion and perihelion. However, the earth moves faster in its orbit at perihelion so the season is shorter, i.e. the number of days between the equinoxes is shorter between September and

March than between March and September. So that the hemispheres and seasons may be compared researchers frequently follow Milankovitch's technique of dividing the year into two equal halves. The division is made so that the total solar receipt every day in one season is less than any day in the other. The resultant seasons are called *caloric summer* and *caloric winter*. As expected, the Northern Hemisphere receives less energy in its caloric summer than the Southern Hemisphere summer. Also the Northern Hemisphere receives more in winter. When the whole year is taken into account, the difference between the hemispheric receipts is small. In fact, (Vernekar, 1972) estimates that the Northern Hemisphere receives more energy annually than the Southern. Inevitably more than one factor must be considered. Thus even if the extremes are used, it has been argued that the temperature contrasts between the hemispheres would be ameliorated by the current distribution of land and sea (the southern hemisphere, which experiences the greatest extremes in receipt, is a water hemisphere).

According to Croll (1864) the first person to consider the effect of eccentricity on climate was Sir John Hershel in 1830. However, estimates of the variation of eccentricity made by Lagrange were suspect because the masses of the minor planets were not accurately known at the time. Therefore, Hershel was unable to reach a positive result. Croll, on the other hand, using contemporary calculations from Leverrier, found that a causal connection between changes in eccentricity and changes in climate was possible. Over a century later Berger (1977) provided more precise estimates of orbital changes, giving $0.0003 < e < 0.0535$ with a significant period in the order of 92,000 years. Other periods are present. These, along with variables of the next sections, have been used in existing theories of climatic change.

Precession

Although no documents survive for him, Hipparchus (2nd century BC) is recognized as being the first to estimate the variation in the alignment of the earth's axis with the "fixed" stars known as the *precession*. Hipparchus was born in Nicaea (Iznik, NW Turkey) but spent his life on the island of Rhodes where he made astronomical observations during the period 147–127 BC. The information on Hipparchus' work comes from Ptolemy. Claudius Ptolemaeus, who lived from ca. AD 100 to ca. AD 170 in Alexandria, is known for a number of written works including the *Almagest*. The title comes from the Latin translation of the Arabic translation of the Greek title. Literally it meant "mathematical [astronomical] compilation." In it Ptolemy writes:

> For Hipparchus too, in his work "On the displacement of the solstical and equinoctial points," adducing lunar eclipses from among those accurately observed by himself, and from those observed earlier by Timocharis, computes

that the distance by which Spica is in advance of the autumnal [equinoctial] point is about 6° in his own time, but was about 8° in Timocharis' time. (Ptolemaeus, 1984, p. 327)

Thus, even in a geocentric universe, Hipparchus had identified precession. At the present time the S–N axis points approximately at the star Polaris whereas in about 13,000 it will point approximately at Vega. This variation has a major period of approximately 22,000 years. Precession affects the time of the equinox. Thus whereas the southern hemisphere summer is now at perihelion, the northern hemisphere summer will be at perihelion in 13,000.

Obliquity

The inclination of the earth's axis to the normal of the plane of the ecliptic is known as the *obliquity*. Currently this is at 23°27′ and is estimated to oscillate from between 22.0° and 24.3° with an important period of approximately 40,000 years (Berger, 1977). The contrast in the incidence of solar radiation between summer and winter increases as the obliquity increases.

6.4.2 Earth–Sun–Space Equilibrium

The Earth as a whole is in approximate equilibrium with regard to energy. That means that over a period of an average year the outgoing radiation (emission) equals the absorption of solar radiation. We ignore the relatively small amounts of energy from stars, reflected from the moon, and being conducted from the earth's interior, etc. (see table 8.1). For an earth without an atmosphere, if we know the fractional absorption, it is relatively easy to calculate the earth's average temperature. We have already seen in section 6.4.1 that the average amount of solar radiation arriving at the top of the atmosphere is $S^* = 1,376\,\text{W m}^{-2}$, the solar constant. Since the earth presents a disc of radius $r = 6,371 \times 10^3$ m perpendicular to the sun's beam, the amount intercepted, I, is

$$intercepted = \pi r^2 S^* \qquad (6.31)$$

The fractional absorption (whose estimate does include the atmosphere) is 0.7. Therefore the average solar energy absorbed is 0.1228×10^{18} W. At equilibrium this energy must be returned to space. If the earth were a black body and the energy were assumed to be emitted uniformly from each square meter of the earth's surface area, $(4\pi r^2 = 510 \times 10^{18}\,\text{m}^2)$, this emission would be

$$emission = \frac{\pi r^2 \times S^* \times 0.7}{4\pi r^2} = \frac{1}{4} S^* \times 0.7 = 240\,\text{W m}^{-2}. \qquad (6.32)$$

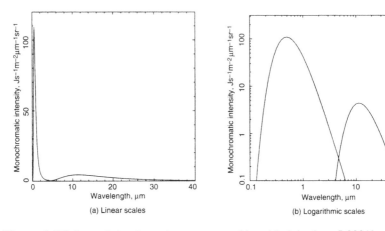

Figure 6.7 Solar emission intensity represented by a black body at 5,880 K adjusted for a distance of 150 × 10⁶ km. Earth emission intensity is represented by a black body at 255 K. Because of the reduction of the solar curve it crosses the earth curve at between 4 and 5 μm instead of completely overlying it (e.g. see the two curves in figure 6.4). 5,880 K represents the theoretical temperature of the sun assuming a solar constant of 1,376 W m⁻² and 255 K represents the theoretical equilibrium temperature of the earth

If we use this in Stefan–Boltzmann's Equation (6.19)

$$\sigma \times T^4 = 240,$$
$$T = 255 \, \text{K}.$$

Since the estimated average temperature of the earth's surface is 288 K other processes are at work. These are collectively known as the greenhouse effect which is described below. At the earth the Planck solar curve (5,880 K reduced for the distance effect) and the earth curve (255 K) overlap at between 4 and 5 μm (see figure 6.7). It so happens that quartz (glass) is a filter at that wavelength, being transparent for shorter wavelengths and being opaque for longer ones.

6.4.3 Trigonometry of Solar Radiation

In the previous section the angle of incidence of solar radiation was ignored because the earth was considered as a whole and energy fluxes were averaged. The angle of incidence of direct solar radiation at the top of the atmosphere or on the earth's surface is one of the most important variables in climatology. The amount of radiation received per unit area is proportional to the cosine of the zenith angle (see figure 6.1). In addition, the zenith angle affects the optical path length (see Equation 6.27). This angle, θ, the zenith angle, or its complement, the altitude angle, α, is a function of the time of day, time of year and latitude.

Time

Since time is so important for many calculations involving the sun its accurate determination is a necessary prerequisite. *Apparent* (true) solar noon occurs when the sun is in the zenith over the meridian of a place, P, but this seldom coincides with clock noon. The relationship between the two depends upon the difference between the meridian of P and the standard meridian, lm; whether daylight savings time is in operation, 1 hour; and the variation of the alignment of the earth with the sun, known as the equation of time, *et*.

The position of the sun is given by the hour angle, H, in degrees from the meridian line, i.e. when converted to time by

$$H_t = \frac{H°}{15°/\text{hour}}.$$ (6.33)

it is the number of hours from apparent solar noon. Subtracted from 12:00 it gives the morning apparent solar time, and added to 12:00 it gives the afternoon apparent solar time in the 24 hour system.

Let the standard meridian be $Long_s$ and the meridian of P be $Long_P$. Here we consider western longitudes negative and eastern longitudes positive. Then

$$lm = \frac{(Long_P - Long_s)°}{15°/\text{hour}}.$$ (6.34)

The equation of time may be obtained from astronomical tables or calculated from algorithms provided by (Meeus, 1991).

Now the conversions from apparent solar time H_t to clock time are given by

$$clock\ time\ (morning) = 12:00 - H_t - et - lm + 1\ \text{hour},$$ (6.35)
$$clock\ time\ (afternoon) = 12:00 + H_t - et - lm + 1\ \text{hour},$$ (6.36)

where the 1 hour is included for daylight savings.

Position of the Sun

Let us consider a location P in figure 6.8 with the vector PZ pointing towards the zenith, vector PS pointing towards the sun and vector PN being the normal to the slope. The latitude is taken to be "L" degrees with northern hemisphere locations assumed positive and southern hemisphere negative. The time of year is represented by the declination of the sun, "δ." This may be obtained from astronomical tables such as List (1958) or calculated from the formulae

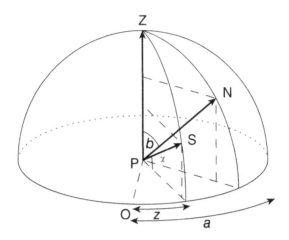

Figure 6.8 Geometry of direct solar radiation on a sloping surface

of Meeus (1991). Time of day is given by the hour angle, "H," which, as indicated in the last section, is measured from noon in degrees.

Positive azimuth angles are measured counterclockwise from south (the line PO). Therefore, morning azimuths, "z," of the sun are positive and east of south and afternoon azimuths are negative and west of south. Similarly eastward facing slopes have positive azimuths, "a."

The altitude angle of the sun is "α" and the zenith angle of the normal to the slope is "b." The angle between the vectors S and N is taken to be "i" and is not labeled in figure 6.8.

The various angular relationships are obtained from simple trigonometry (Milankovitch, 1969; Brooks, 1960).

The altitude of the sun is given by

$$\sin(\alpha) = \cos(L)\cos(\delta)\cos(H) + \sin(L)\sin(\delta), \qquad (6.37)$$

and the azimuth by

$$\cot(z) = \frac{\sin(L)\cos(H) - \cos(L)\sin(\delta)}{\sin(H)}, \qquad (6.38)$$

or

$$\sin(z) = \frac{\cos(\delta)\sin(H)}{\cos(\alpha)}. \qquad (6.39)$$

Sunrise and sunset are said to occur when the upper limb of the sun's disc is just on the horizon. Since solar position in the above equations refers to the center of the sun, sunrise and sunset occur when the sun's center is below the horizon by the semi-diameter which is 16 seconds of angle. Also, due to refraction, the upper rim of the sun is actually about 34 seconds (variable) below the horizon at that time. Combined these make α_0 at sunrise and sunset -50 seconds $= -0.8333°$ (Meeus, 1991). From Equation (6.37)

$$\cos(H_0) = \frac{\sin(\alpha_0) - \sin(L)\sin(\delta)}{\cos(L)\cos(\delta)}. \tag{6.40}$$

Sloping Surfaces

Now the angle "i," enclosed by NPS, is given by

$$\cos(i) = \sin(b)\cos(\alpha)\cos(z - a) + \cos(b)\sin(\alpha) \tag{6.41}$$

For special cases Equation (6.41) may be written in alternative forms. For a horizontal surface

$$\cos(i) = \sin(\alpha). \tag{6.42}$$

For a north to south oriented slope

$$\cos(i) = \cos(L - b)\cos(\delta)\cos(H) + \sin(L - b)\sin(\delta), \tag{6.43}$$

or

$$\cos(i) = \sin(b)\cos(\alpha)\sin(z) + \cos(b)\sin(\alpha). \tag{6.44}$$

For vertical walls

$$\cos(i) = \cos(\alpha)\cos(z - b). \tag{6.45}$$

For east facing slopes

$$\cos(i) = \cos(L - b)\cos(\delta)\cos(H) + \sin(L - b)\sin(\delta). \tag{6.46}$$

6.4.4 Atmospheric Gases

As indicated in section 6.2, gases selectively absorb and emit: the full spectrum is a signature of the substance. Figure 6.9 displays very generalized absorption spectra obtained in the laboratory for some of the more important atmospheric gases. Also included for comparison is the solar spectrum measured at the surface. There is clearly a subjective correlation showing the combined effect of these gases. Given the absorption spectrum and the optical depth of the particular gas, absorption is a function of the incident wavelengths while ideal emission is a function of the temperature of the gas as specified by Planck's Law.

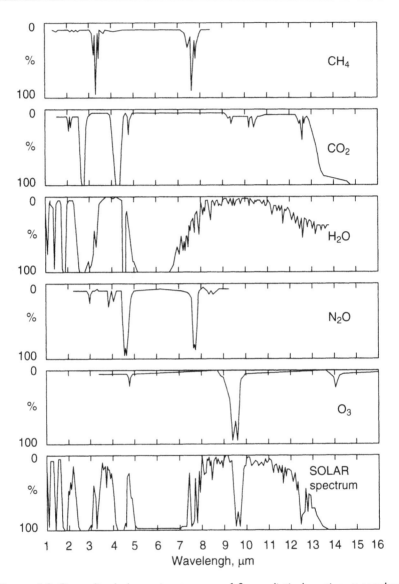

Figure 6.9 Generalized absorption spectra of five radiatively active atmospheric gases and the solar spectrum at sea level. Data modified from Howard (1965)

In principle, Equation (6.29) may be integrated over all wavelengths, over all directions (both up and down), over all absorbers, over all layers. Because of the fine detail in the absorption spectra and the variation in absorbing gases, especially in water vapor, this turns out to be a most complex task.

Initially estimates of long wave radiation were made using graphical methods (Elsasser, 1942; Charney, 1945; Yamamoto, 1952) but now they are performed digitally.

Of the solar radiation that enters the top of the atmosphere (usually considered to be wavelengths 0.1 to 5 μm) only about one-fifth is absorbed by the atmosphere whereas about half is absorbed by the surface. The rest is returned to space in the same wavelengths as they arrive. This amounts to 30 percent. Therefore the *albedo*, the reflectivity in short wavelengths, of the earth (and atmosphere) is said to be 0.30.

The vertical temperature profile in the troposphere is maintained by the combination of radiation and vertical motions involving adiabatic processes and latent heat release. The resulting average lapse rate is $6.5°C\,km^{-1}$. Most of the radiation emitted by the earth (wavelengths 5 to 30 μm) is absorbed and re-emitted by the atmospheric "greenhouse gases." Much of it returns to the surface but, as dictated by the conservation of energy, an amount equivalent to that absorbed from the sun is emitted to space. However, most of this emanates from the higher atmosphere where temperatures are much lower than the surface. Figure 6.10 shows a satellite receipt of long-wave radiation from the earth–atmosphere from Guam. The atmospheric window region, 8 to 14 μm, is composed mainly of emission from the ground which is in the order of 20°C. A band in the center of that region is emitted by ozone at temperatures less than 0°C and a broad band 14 to 16°C is due to carbon dioxide at about −60°C. The average temperature of the combined emission, assuming a black body system, will approach the theoretical 255 K.

Thus, while the equilibrium temperature of the earth is 255 K, a much higher (by 33 K) temperature at the surface is maintained. The whole process is called the *greenhouse effect* even though most people refer only to the radiative aspects (glass transparent to solar, opaque to long waves) and even though the analogy with the greenhouse is not a good one (glass inhibits vertical mixing).

The earliest reference to the process whereby the atmospheric gases allow the transmission of solar radiation but inhibit the direct emission of earth radiation to space is by Fourier (1824). He used a "glass bowl" as an analogy (Rodhe et al., 1997).

Later John Tyndall popularized the concept. Tyndall (1820–93) was born and educated in Ireland. He became a surveyor and engineer and moved to England where he later (1847) taught mathematics and drawing at a school in Hampshire. There he met Edward Frankland and with him went to the University of Marburg, Germany, where he completed a doctorate. He was unable to obtain a job in England so, like his friend T. H. Huxley, he began writing and lecturing to popularize science. With the assistance of Faraday, Tyndall became a professor at the Royal Institution in 1853.

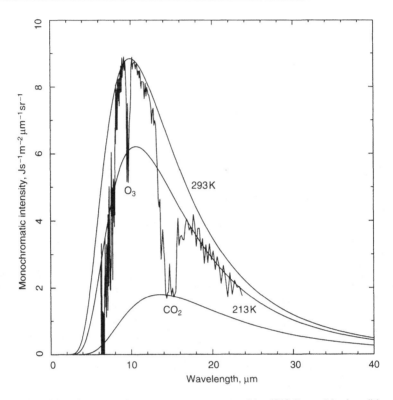

Figure 6.10 Earth–atmosphere emission measured by IRIS-D on Nimbus IV near Guam at 15.1°N on April 27, 1970 with black body curves 213, 273, and 293 K. Data modified from Kunde et al. (1974)

For Tyndall the "atmospheric effect" was due to water vapor. In 1861 in "The Bakerian Lecture" delivered before the Royal Society, Tyndall (1961) stated: "Hence the differential action of the heat coming from the sun to the earth into space, is vastly augmented by the aqueous vapour of the atmosphere." Tyndall also recognized that changes in the amount of greenhouse gases could change climate. He had thus identified what we now call the *enhanced green-house effect*. In a commentary to the previously quoted memoir, published in 1873, he wrote:

it is here pointed out that a comparatively slight change in the variable con-stituents of our atmosphere, by permitting free access of solar heat to the earth, and checking the outflow of terrestrial heat towards space, would pro-duce changes of climate as great as those which the discoveries of geology reveal. (Tyndall, 1873, p. 4)

Tyndall did not use the greenhouse analogy. Instead he used a river dam:

> As a dam built across a river causes a local deepening of the stream, so our atmosphere, thrown as a barrier across the terrestrial rays, produces a local heightening of the temperature at the earth's surface. This, of course, does not imply indefinite accumulation, any more than the river dam does, the quantity lost by terrestrial radiation being, finally, equal to the quantity received from the sun. The chief intercepting substance is aqueous vapour of the atmosphere . . . (Tyndall, 1873, p. 117)

The first quantifier of the greenhouse effect as produced by carbon dioxide is Svante Arrhenius (1896). Arrhenius (1859–1927) graduated with a doctorate from Uppsala in 1884 and became a teacher of physics at the precursor of the University of Stockholm in 1891. His original intention in his atmospheric work was to explain the temperature variation in glacial cycles but he extended this to look at possible future changes. A full review of his work and recent findings occupy the February 1997 issue of The Royal Swedish Academy of Sciences Journal, *Ambio*.

The term "hothouse" was used in the nineteenth century but according to Safire (1990) the term "greenhouse effect" was introduced by the Wisconsin

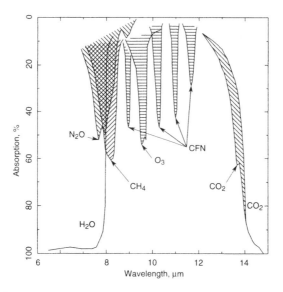

Figure 6.11 Generalized absorption in the infrared window region. Additional CO_2 causes an expansion of absorption into lower wavelengths in the 13–14 μm range which are shaded. Data modified from figure 11.12 of Turco (1995)

geography Professor Glenn T. Trewartha in 1937 in the *Oxford English Dictionary Supplement.*

In recent decades a large amount of research has been conducted into the enhanced greenhouse effect especially as it relates to *global warming*. Not only are existing constituents of the atmosphere increased, new ones are introduced. Figure 6.11 displays the influence of these on absorption in the window region.

6.5 Comment

Again we have artificially isolated a segment of knowledge which relates to the atmosphere. Clearly solar radiation and its selective absorption is the energy source that drives the whole system. It is also the mechanism whereby that energy is eventually dispersed to space. Radiation then is an integral and central component of dynamic climatology.

Chapter 7

Atmospheric Equations

7.1 The Nature of Fluids

So far we have considered the concepts of mechanics applied to solid objects. The atmosphere is a fluid. Fluids encompass both liquids and gases. On a simplistic level it might appear that the distinction between solids and fluids is that solids have shapes that are not easily changed and fluids in bulk have no predetermined shape and deform very easily. In fact, there is a continuum and some substances may have both characteristics. However, for our purposes, where we are only dealing with air and water, the above distinction is sufficient.

The difference between liquid and gas is mainly one of density. Also gases may be compressed much more easily than liquids. At the molecular level the molecules are arranged in an ordered way in a solid, partially ordered in a liquid and disordered in a gas (see section 5.2.8). We have been concerned with the molecular scale when dealing with micro-scale view of heat and radiation but for the most part we consider some unit mass (one kilogram) or volume (one cubic meter) of a fluid containing large numbers of molecules. Then, even when we are considering a finite volume having six plane sides, we assign the mass of the element to its center of gravity.

A basic assumption here is that mass is conserved: it cannot be created or destroyed although more or less of it may be concentrated in a constant volume. For a unit mass element, the acceleration, also called the *inertial force*, is affected by two types of forces that have already been introduced in section 4.3.7. The long range forces are due to accelerations of gravity and curved motion and the short range forces are due to the stresses of friction and pressure. Friction is the name given to the force which opposes relative motion. In a fluid it is proportional to *viscosity*. In completely general terms, friction in a fluid must take into account: 1) the random motion of molecules; and 2) the random motion of fluid elements, which is called *turbulence*. Since the theory of the former was developed and incorporated into the equations of motion early it will be considered first in the following.

7.2 Continuity – Conservation of Mass

The conservation of mass in a fluid was elaborated by Euler between 1753 and 1755 (Youschkevitch, 1981).

Leonhard Euler (1707–83) was born in Basel, Switzerland, where he entered the university in 1720. Despite being deeply religious and initially wanting to follow in his father's footsteps and become a protestant minister, he soon became interested in mathematics. An early influence was Johann Bernoulli whose sons Nikolaus II and Daniel went to Russia. That connection led to Euler being appointed to the one-year-old St. Petersburg Academy of Sciences in 1726 as an adjunct of physiology. He became professor of physics in 1731 and succeeded Daniel as professor of mathematics in 1733. In 1740 the political situation in Russia changed and Euler accepted an appointment from Frederick the Great in Berlin in 1741. However, he continued to work with the St. Petersburg Academy. He returned permanently to Russia in 1766.

Leonhard had a phenomenal memory and was an inexhaustible researcher. He was always slow in finalizing his publications, however, and his ideas were often made known before they were formally printed. In the first 14 years in St. Petersburg he prepared between 80 and 90 works of which only 55 were published. During his 15 years in Berlin he prepared 380 and published about 275. Many of his discoveries in mathematics and mechanics were brilliant and, as mentioned earlier, are still current today. In 1733 he worked on maps in the department of geography.

In fluid mechanics we recognize especially Euler's development of his general equations and the conservation of mass equation but his influence in the field goes well beyond these. The conservation of mass equation, or as it is usually known, the *equation of continuity* may be given in various forms. As shown by Lamb (1932) following Euler, the Cartesian versions of these may be derived from the basic relationship that an element of mass, M, which is the product of density, ρ, and volume, $\mathcal{V} = \delta x \, \delta y \, \delta z$, cannot change with time

$$\frac{d(\rho \mathcal{V})}{dt} = 0, \tag{7.1}$$

or,

$$\frac{\partial \rho}{\partial t} + \nabla \cdot (\rho \mathbf{V}) = 0, \tag{7.2}$$

or,

$$\frac{1}{\rho}\frac{d\rho}{dt} + \nabla \cdot \mathbf{V} = 0, \tag{7.3}$$

or,

$$\frac{1}{\alpha}\frac{d\alpha}{dt} - \nabla \cdot \mathbf{V} = 0. \tag{7.4}$$

For an incompressible fluid, $d\rho/dt = 0$, and

$$\nabla \cdot \mathbf{V} = 0. \tag{7.5}$$

Lamb, however, did not use vectors. Sir Horace Lamb (1849–1934) came from a broken family in Stockport, England. His father died when he was two and his mother was abused by her second husband. Consequently he was brought up by an aunt. Lamb's unusual abilities were recognized by a schoolmaster who encouraged him to switch from classics to mathematics and the 17 year-old won a mathematics scholarship to Trinity College, Cambridge in 1867. There he had an outstanding record. As stated by his grandson, the climatologist, Hubert Lamb,

> My father's father was old Horace Lamb, renowned mathematician, whose pioneering book on hydrodynamics, originally drafted from his lecturing notes in 1879, was reprinted many times and still brought in a small income to his descendants until after 1990, an almost unheard-of longevity for a scientific textbook. (Lamb, 1997)

The hydrodynamics book remains a classic and the 1932 edition has been used frequently by this author.

7.3 Molecular Viscosity

The molecular aspects of friction are based upon the concept of laminar flow, by which we mean that the fluid moves smoothly in parallel layers or sheets. Because there is no turbulence, the only interchange between layers is by the molecules themselves. The effect of the interchange is the establishment of the forces known as the *viscous stress*, or *shearing stresses*. Such forces transfer momentum and convert the kinetic energy of the bulk motion of the fluid to the molecular motion of translation (temperature-heat). Analysis of the molecular viscous stresses may be approached either from macro-scale concepts (Newtonian method) or from micro-scale concepts (kinetic theory of gases).

7.3.1 Newtonian Approach

Newton (1687) in Book II Section IX, states:

> Hypothesis. The resistance arising from the want of lubricity in the parts of a fluid, is, other things being equal, proportional to the velocity with which the parts of the fluid are separated from one another.

Today this may be developed in the following way. Imagine two plates held parallel and separated a small distance by a fluid. Here the arrangement is assumed to be in the xz plane but it could be in any plane. Experimentally it may be shown that, if the bottom plate is held rigidly, a constant force must be exerted on the upper one to keep it moving at a small and constant speed. The force, $\mathbf{F}$, in turn, is proportional to: 1) the area of the upper plate, (a, m^2); 2) the speed $(u, \text{m s}^{-1})$; and inversely proportional to 3) the distance (z, m) separating the plates i.e.

$$F = \mu \frac{au}{z}, \text{kg m s}^{-2} \tag{7.6}$$

where μ is the constant of proportionality and is known as the *molecular* or *dynamic* viscosity, or more fully as the *coefficient of dynamic viscosity*. It has units of $\text{kg m}^{-1}\,\text{s}^{-1}$. Observation also shows that the velocity changes linearly between the plates as shown in figure 7.1. Therefore

$$\frac{\partial u}{\partial z} = \frac{u}{z}, \text{s}^{-1},$$

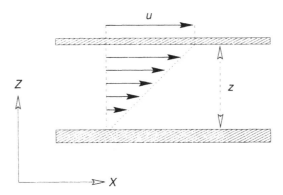

Figure 7.1 Friction due to random molecular motion. The upper plate is moved with a constant force $\mathbf{F}$ at a constant speed u while the lower plate is fixed. The intervening fluid is assumed to exhibit laminar flow

where $\partial u/\partial z$ is called the velocity gradient or shear and $u = f(z)$, the velocity profile. The above equation may be used to replace u/z in Equation (7.6). Furthermore, if the force is written in terms of unit area, the shearing stress, τ, is given by

$$\tau = \frac{F}{a} = \mu \frac{\partial u}{\partial z} = \nu\rho \frac{\partial u}{\partial z}, \text{ kg m s}^{-2}\text{ m}^{-2}, \tag{7.7}$$

where $\nu = \mu/\rho$ is the *kinematic* viscosity in units of m^2 s^{-1}.

7.3.2 Statistical Approach

We shall follow the statistical mechanics approach that molecules are approximately spherical with definite diameters and that they move around randomly at an average translational velocity which is dependent upon temperature. Their characteristics for the atmosphere near sea level are given in section 5.2.8. The random motions are superimposed upon the average flow which is observed as the ordered motion of the fluid. When molecules collide, momentum (Mu) is transferred between them, thus speeding up slower moving ones and slowing down the faster ones. In the case where a velocity profile exists, the random motion will cause an interchange of molecules having different average ordered motion thus leading to a transfer between layers.

Assume that an individual molecule of mass, $M(z)$ started its vertical motion at some level z with horizontal velocity $u(z)$ and that it travels a distance l before it collides with another molecule $M(z + l)$ of velocity $u(z + l)$. The transfer of momentum by the molecule is therefore

$$M(z + l)u(z + l) - M(z)u(z)$$

The velocity $u(z)$ may be expanded as a power series, i.e.

$$u(z) = u_0 + \frac{\partial u}{\partial z}l + \frac{1}{2}\frac{\partial^2 u}{\partial z^2}l^2 + \cdots$$

Now, in laminar flow the velocity profile is linear. Also, the molecules are assumed to be of uniform size. Therefore the momentum transfer for one molecule may be written

$$Ml\frac{\partial u}{\partial z}.$$

The total transfer by all molecules will depend upon their velocity c and the number per unit volume which travel in the vertical direction. If N is the total number then $\frac{1}{3}N$ may be assumed to move vertically. Consequently the total transfer will be

$$\frac{1}{3} N M c l \frac{\partial u}{\partial z}$$

where $NM = \rho$, the density. This transfer of momentum in a layer is equivalent to a force, the shearing stress. Therefore

$$\tau = \frac{1}{3} \rho c l \frac{\partial u}{\partial z} \tag{7.8}$$

where

$$\frac{1}{3} \rho c l = \mu,$$

and

$$\frac{1}{3} c l = \nu,$$

leading, for example, to

$$\tau = \rho \nu \frac{\partial u}{\partial z}. \tag{7.9}$$

As pointed out by Sutton (1953) the above argument is extremely crude yet it is sufficient for our purposes. A number of comments are pertinent at this time: 1) The net transfer of molecules is zero. Otherwise there would be a transfer of mass. 2) The same argument may be applied to the other coordinates. 3) The stress is a function of the agitation (temperature) and the organized velocity gradient. Estimates for μ and ν from Sutton (1953) and Weast (1965) are given in table 7.1.

Notice that the kinetic viscosity is much greater for air than for water so that momentum is diffused much more rapidly in air.

Table 7.1 Dynamic and kinematic viscosities (μ: kg m^{-1} s^{-1}; ν: m^2 s^{-1})

	°C	−9.3	0.0	20.0	40.0	
Atmos	μ		0.0171	0.0181	0.0190	$\times 10^{-3}$
	ν		0.0132	0.0150	0.0169	$\times 10^{-3}$
Water	μ	2.549	1.798	1.002	0.654	$\times 10^{-3}$
	ν	2.549	1.798	1.004	0.00659	$\times 10^{-6}$

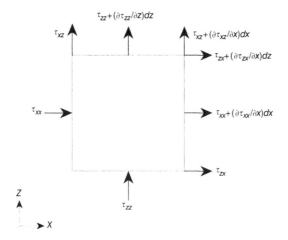

Figure 7.2 Stress tensor in the xz plane. Of the nine possible, four components plus the first terms of their increments across the element are shown

Sir (Oliver) Graham Sutton (1903–77) is widely known for his 1953 text on Micro-meteorology and, as the first Director General of the British Meteorological office 1953–65, he was responsible for acquiring the first major computer for that office in 1958. He was a mathematician having studied under G. H. Hardy at Oxford.

7.4 The Stress Tensor

The stress per unit area, τ, as discussed above, is only one of nine stresses which may be derived for an element of fluid. For example, figure 7.2 shows all the stresses acting in the x and z directions on the faces perpendicular to x and z. The first subscript indicates the face on which the stress acts (e.g. τ_{xz} acts on the face perpendicular to x) and the second subscript its direction (e.g. τ_{xz} acts in the z direction). Note that higher order derivatives of the Taylor expansions have been ignored in the increment across the element.

The whole set of stresses, when written as a matrix, is known as a stress tensor:

$$\begin{matrix} \tau_{xx} & \tau_{xy} & \tau_{xz} \\ \tau_{yx} & \tau_{yy} & \tau_{yz} \\ \tau_{zx} & \tau_{zy} & \tau_{zz} \end{matrix}$$

The diagonal elements of this tensor are known as the normal stresses and those off the diagonals the tangential stresses. As shown by Lamb (1932), the normal stresses are:

$$\tau_{xx} = -p - \frac{2}{3}\mu \left(\frac{\partial u}{\partial x} + \frac{\partial v}{\partial y} + \frac{\partial w}{\partial z} \right) + 2\mu \frac{\partial u}{\partial x},$$

$$\tau_{yy} = -p - \frac{2}{3}\mu \left(\frac{\partial u}{\partial x} + \frac{\partial v}{\partial y} + \frac{\partial w}{\partial z} \right) + 2\mu \frac{\partial v}{\partial x}, \qquad (7.10)$$

$$\tau_{zz} = -p - \frac{2}{3}\mu \left(\frac{\partial u}{\partial x} + \frac{\partial v}{\partial y} + \frac{\partial w}{\partial z} \right) + 2\mu \frac{\partial w}{\partial x},$$

and the tangential stresses are:

$$\tau_{yz} = \mu \left(\frac{\partial w}{\partial y} + \frac{\partial v}{\partial z} \right) = \tau_{zy},$$

$$\tau_{zx} = \mu \left(\frac{\partial w}{\partial y} + \frac{\partial v}{\partial z} \right) = \tau_{xz}, \qquad (7.11)$$

$$\tau_{xy} = \mu \left(\frac{\partial w}{\partial y} + \frac{\partial v}{\partial z} \right) = \tau_{yx}.$$

The local (short range) forces present in a fluid are given by the gradients of the stresses with respect to distance. The form of these forces had been developed earlier and incorporated into the equations of motion in a fluid as outlined in the next section.

7.5 Navier–Stokes Equations

Whereas the equations describing the motion of solid bodies were set down by Newton in 1665, those for fluids had to wait more than 180 years. It was Stokes (1845) who published a full derivation of the partial differential equations applied to fluids in motion. In that paper he acknowledged the work of Poisson (1829) stating "that Poisson had written a memoir on the same subject, and on referring to it I found that he had arrived at the same equations." As a footnote Stokes also indicated that Navier (1822) had obtained the same equations in the case of an incompressible fluid. However, as pointed out by Lamb (1932) Art. 328, both Navier and Poisson had used various considerations as to the mutual action of the ultimate molecules of fluids. Stokes did not rely upon this approach. de Saint-Venant (1843) also apparently independently arrived at these results (Lamb, 1932). It is Navier alone, however, who has his name recognized along with Stokes for these basic equations.

Claude-Louis-Marie-Henri Navier (1785–1836) studied in Paris under his mother's uncle, an engineer, for entrance to École Polytechnique in 1802. There he became a protégé of Fourier although he moved to the École des Ponts et Chaussés graduating in 1806. He taught there until 1831 when he replaced Cauchy at École Polytechnique. He worked in engineering and on problems of mechanics overlapping the work of Carnot and Coriolis. In his derivation of partial differential equations for the motion of liquids he applied Fourier methods to find particular solutions. However, he had no concept of shear and his ideas on molecular interaction would be unacceptable today (McKeon, 1981).

George Gabriel Stokes (1819–1903), born in Skreen, Ireland, was tutored by his rector father before attending Bristol College in preparation for Pembroke College, Cambridge. There he excelled in mathematics and upon graduating in 1841 was immediately appointed Lucasian professor, Newton's old title. Unfortunately the chair's endowment had deteriorated and Stokes had to augment his income by teaching at the Government School of Mines in London. Although he is now known for his fluid work, Stokes was a recognized geodesist and expert in optics.

Stokes claimed that he was not aware of the French work until after he had written his 1845 paper and that, anyway, his assumptions were different (Stokes, 1845). His equation (12) is

$$\rho \left(\frac{Du}{Dt} - X \right) + \frac{dp}{dx} - \mu \left(\frac{d^2u}{dx^2} + \frac{d^2u}{dy^2} + \frac{d^2u}{dz^2} \right)$$
$$- \frac{\mu}{3} \frac{d}{dx} \left(\frac{du}{dx} + \frac{dv}{dy} + \frac{dw}{dz} \right) = 0, \& c \cdots \qquad (7.12)$$

Of course, Stokes' lower case ds are partials. Together with similar equations for the other components the set is known as the *Navier–Stokes Equations*. They are the starting point for texts in fluid dynamics. Sometimes they are introduced in modified form. For example, if the fluid is assumed to be incompressible then, from Equation (7.5), the last term on the left becomes zero. If, in addition, the fluid is inviscid, $\mu = 0$, and the third term becomes zero, resulting in what are known as *Euler's Equations of Motion*.

7.6 Turbulent Eddy Viscosity

So far we have considered a non-turbulent fluid. Once laminar flow breaks down, bulk elements of the fluid will move between layers. These elements must transfer momentum very much like the molecules in non-turbulent flow. The first person to investigate this phenomenon was Osbourne Reynolds (1883). Reynolds (1842–1912), born in Belfast, was apprentice to a mechanical engineer before attending Cambridge and graduating in mathematics in 1867. The following year he became professor of engineering at Owens College in Manchester, where he stayed for 37 years, focusing on the mechanics of fluids including the mechanical equivalent of heat.

An important contribution of Reynolds' research was the identification of a non-dimensional variable that defined the characteristics necessary for a change from laminar flow. This was originally defined by him for water in a pipe as

$$\rho \frac{D\,U_m}{\mu} = K,\tag{7.13}$$

where D was the diameter of the pipe and U_m was the mean speed of the water.

This was generalized as the ratio of the inertial force to the viscous force in fluid motion and is known as the *Reynolds number*,

$$Re = \frac{LU}{\nu},\tag{7.14}$$

where L is a characteristic length and U is a *characteristic* velocity. (For a discussion of characteristic quantities see section 7.9.)

7.6.1 Reynolds' Stresses

Reynolds' approach to estimating turbulent eddy viscosity depended upon using the motion of the fluid elements themselves. As an example, the vertical transfer by the vertical component wind of the horizontal u component of momentum is the product $Mu \times w$. In fact, there are three components of momentum and three components of wind to effect the transfer.

Also, the variables, say u and w, may be decomposed into their mean or average, represented by the overbar, and deviation from the mean, represented by the prime, i.e.

$$u = \overline{u} + u'\tag{7.15}$$
$$w = \overline{w} + w'\tag{7.16}$$

Multiply, and we have

$$uw = \overline{u}\,\overline{w} + u'w' + \overline{u}w' + u'\overline{w}.$$

Average, recognizing that the average of a constant times the deviation from the average is zero, and we obtain

$$\overline{uw} = \overline{u}\,\overline{w} + \overline{u'w'}.\tag{7.17}$$

This process is known as *Reynolds' resolution*.

If uw is multiplied by density the result is the flux of momentum or force per unit area, now known as the Reynolds' stress. Thus, by applying such relationships Reynolds was able to expand Navier–Stokes equations to include the turbulent elements, (Reynolds, 1894, Equation 15)

$$\rho \frac{\partial \overline{u}}{\partial t} = \frac{\partial}{\partial x}\left(\overline{\tau}_{xx} - \rho \overline{u}^2 - \rho \overline{u'^2}\right) + \frac{\partial}{\partial y}\left(\overline{\tau}_{xy} - \rho \overline{u}\,\overline{v} - \rho \overline{u'v'}\right)$$
$$+ \frac{\partial}{\partial z}\left(\overline{\tau}_{xz} - \rho \overline{u}\,\overline{w} - \rho \overline{u'w'}\right)$$

$$\rho \frac{\partial \overline{v}}{\partial t} = \frac{\partial}{\partial x}(\overline{\tau}_{xy} - \rho \overline{u}\,\overline{v} - \rho \overline{u'v'}) + \frac{\partial}{\partial y}(\overline{\tau}_{yy} - \rho \overline{v}^2 - \rho \overline{v'^2})$$
$$+ \frac{\partial}{\partial z}(\overline{\tau}_{yz} - \rho \overline{v}\,\overline{w} - \rho \overline{v'w'})$$

$$\rho \frac{\partial \overline{w}}{\partial t} = \frac{\partial}{\partial x}(\overline{\tau}_{xz} - \rho \overline{u}\,\overline{w} - \rho \overline{u'v'}) + \frac{\partial}{\partial y}(\overline{\tau}_{yz} - \rho \overline{v}\,\overline{w} - \rho \overline{v'w'})$$
$$+ \frac{\partial}{\partial z}(\overline{\tau}_{zz} - \rho \overline{w}^2 - \rho \overline{w'^2}). \tag{7.18}$$

7.6.2 Molecular Analogy

Using similar arguments as in section 7.3 we may derive a relationship analogous to Equation (7.9) for the vertical:

$$\rho \, K_m \frac{\partial \overline{u}}{\partial z}. \tag{7.19}$$

Here the parcel is assumed to have originated at a level having a mean horizontal velocity $\overline{u}$ and the coefficient K_m is known as the *coefficient of eddy viscosity*. Earlier, the quantity $A = \rho K_m$, named the *Austausch* (exchange) coefficient by Schmidt (1925), was used. The assumptions underlying this equation are less valid than for the earlier one. For example, molecular viscosity is a function of temperature, turbulent viscosity is a function of non-linear temperature and the wind gradients which are themselves not related to one another in a simple way. This is especially true of the vertical components. As a result the eddy viscosity is a difficult parameter to estimate.

The total shearing stress equivalent to Equation (7.9) may now be written as a combination of the two viscosities as

$$\tau = \rho(\nu + K_m)\frac{\partial \overline{u}}{\partial z}. \tag{7.20}$$

The first application of turbulence theory to the atmosphere is due to Taylor (1915). Spurred by the Titanic disaster, Geoffrey Ingram Taylor (1886–1975), Reader of Dynamic Meteorology at Cambridge, made measurements from the whaler, Scotia, off Newfoundland in 1913. From these he applied Reynolds' techniques to estimate the eddy viscosity and the vertical transfers of momentum and heat. He found K_m to be in the order of $0.05\,\mathrm{m^2\,s^{-1}}$ which is much larger than the kinematic viscosity, ν (see table 7.1). Consequently ν is usually ignored, and reference to Navier–Stokes Equations in atmospheric work is less frequent than in general fluid dynamics. Also, friction usually appears as the symbol, **F**.

7.6.3 Extension to Other Fluxes

Because eddies are involved in the fluxes of other quantities such as heat and moisture it is appropriate to discuss the relevant equations here. Although horizontal transfers also take place they are usually ignored. If we use the direct approach of the Reynolds resolution as in Equation (7.18) we get for the three fluxes

$$\tau = \overline{\rho w u} \approx \rho \overline{w'u'},$$

$$Q_H = c_p \overline{\rho w \theta} \approx c_p \rho \overline{w'\theta'},$$

and

$$E = \overline{\rho w q} \approx \rho \overline{w'q'}, \tag{7.21}$$

where Q_H is *sensible heat* flow, θ is potential temperature, and E is water vapor flux. The approximate results are obtained from the assumptions that $\rho = constant$ and $\overline{w}\,\overline{s} = 0$ for the averaging period. These equations use covariances or correlations so this method for representing the fluxes is often referred to as the *eddy correlation* approach.

On the other hand if we use Equation (7.19) as the standard form we get

$$\tau = \rho K_m \frac{\partial \overline{u}}{\partial z},$$

$$Q_H = \rho K_H c_p \frac{\partial \overline{\theta}}{\partial z},$$

and

$$E = \rho K_E \frac{\partial \overline{q}}{\partial z}, \tag{7.22}$$

where K_H and K_E are known as *eddy diffusivities* in units of $\mathrm{m^2\,s^{-1}}$. They have magnitudes between 10^{-2} and 10^2. Like the eddy viscosity they characterize the amount of turbulent mixing taking place.

The equation set (7.22), which typically includes the wind profile, is known as the *aerodynamic* approach.

7.6.4 Richardson Number

Whereas superficially it may appear that the Ks in Equation (7.22) should be equal, closer examination reveals otherwise. Vertical mixing (turbulence) is produced by two mechanisms: buoyancy, known as *thermal* turbulence, and wind shear, known as *mechanical* turbulence. Therefore, the vertical motion w will be a function both of both the potential temperature and the gradient of the horizontal wind in the vertical. Also, the Ks, whose magnitudes are sought, will be functions of these turbulence generating mechanisms. It is useful, therefore, to have some measure of the relative magnitudes of the two processes. Such an indicator was developed by Richardson (1920), which he described as a criterion for "just no turbulence," but which is now known as the *Richardson number*, R_i. Sometimes this characteristic is given as R_f, the *Richardson flux number*:

$$R_i = \frac{K_m}{K_H} R_f = \frac{(g/\hat{\theta})(\partial \overline{\theta}/\partial z)}{(\partial \overline{u}/\partial z)^2}, \tag{7.23}$$

where $\hat{\theta}$ is the average potential temperature for the layer, and $\overline{\theta}$ is the average at a specific height. For neutral conditions ($\overline{\theta}$ constant with height) R is zero, for stable or inversion conditions ($\overline{\theta}$ increases with height) R is positive, and for lapse conditions R is negative. Together with the Reynolds number the Richardson number is a basic parameter of turbulence.

7.7 The Vector Equation of Motion

7.7.1 Relative Acceleration

In atmospheric work the Navier–Stokes Equations are typically derived from Newton's Second Law in terms of unit mass, i.e.

$$\left(\frac{d\mathbf{V}_I}{dt}\right)_{Inertial} = \text{Forces per unit mass} = \text{acceleration}. \tag{7.24}$$

The subscript *Inertial* emphasizes that the equation applies to an inertial frame of reference, a fact of which Newton was well aware. By that we mean that the coordinate system is assumed to be non-accelerating. For atmospheric flow it is typically assumed that a frame located at the center of the earth is inertial: the effect of earth revolution is small enough to be ignored. This still leaves earth rotation within Equation (7.24). Because we like to consider acceleration relative to our position on the earth's surface the rotation effect is separated out. From Equation (2.61)

$$\left(\frac{d\mathbf{V}_I}{dt}\right)_I = \left(\frac{d\mathbf{V}_I}{dt}\right)_{rel} + \Omega \times \mathbf{V}_I.$$

But $\mathbf{V}_I = \mathbf{V}_{earth} + \mathbf{V}_{rel}$ and $\mathbf{V}_{earth} = \Omega \times \mathbf{r}$ where $\mathbf{r}$ is the vector (distance from earth center). Substitution leads to

$$\left(\frac{d\mathbf{V}_I}{dt}\right)_{Inertial} = \left(\frac{d\mathbf{V}_{rel}}{dt}\right)_{rel} + (2\Omega \times \mathbf{V}_{rel}) + [\Omega \times (\Omega \times \mathbf{r})]. \qquad (7.25)$$

7.7.2 The Complete Vector Equation

The Second Law may now be written

$$\left(\frac{d\mathbf{V}_{rel}}{dt}\right)_{rel} = -(1/\rho)\nabla p - (2\Omega \times \mathbf{V}_{rel}) - \mathbf{f} - [\Omega \times (\Omega \times \mathbf{r})] + \mathbf{F}.$$

It has already been shown in section 4.3.3 that $\mathbf{f} + [\Omega \times (\Omega \times \mathbf{r})] = \mathbf{g}$. Consequently we may now drop the subscripts and assume everything is in a relative coordinate frame of reference. Hence, the usual form of the vector *Equation of Motion* in atmospheric work is:

$$\frac{d\mathbf{V}}{dt} = -(1/\rho)\nabla p - (2\Omega \times \mathbf{V}) - \mathbf{g} + \mathbf{F}. \qquad (7.26)$$

The relative acceleration of the wind is equal to the pressure gradient force per unit mass, the *Coriolis* force per unit mass, the acceleration of gravity and the frictional force per unit mass, all in units of $\mathrm{m\,s^{-2}}$ (see section 4.3.5 "Work" for a note on Gaspard Coriolis).

7.8 General Coordinates

One of the conveniences of vector notation is that it applies to all general coordinates. Unfortunately many people cannot translate the vector equations in their minds to the local environment where the different terms are observed as winds, accelerations, and gradients and where such vectors are applied as finite differences in calculations. The solution is to replace vectors by their components, there is a question concerning which system to select. The simplest is the Cartesian system, but it has limitations on a spherical surface like the earth. The most convenient system for flow that is closely related to the shape of the earth is the *spherical polar coordinate* system. Another useful set of coordinates for rotating bodies is what is called the *natural coordinates* system. These three are standard in most introductory textbooks such as Holton (1979).

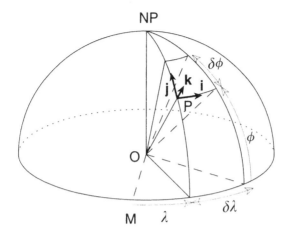

Figure 7.3 The three unit vectors in the atmosphere at point P are functions of longitude λ, latitude ϕ, and height z (not shown). **i** points towards the east (x, λ), **j** towards the north (y, ϕ) and **k** towards the zenith (z). M is the prime meridian. Each vector is of the same length but may appear otherwise because of the perspective. In this section each vector in turn will be moved through small, but exaggerated steps in the following diagrams, distances δx $(r \cos \phi \, \delta \lambda)$, δy $(r \, \delta \phi)$, and δz, respectively

7.8.1 Cartesian Coordinates

Following the procedure in Equation (2.52) we may expand Equation (7.26). Then the local acceleration term is:

$$\frac{d\mathbf{V}}{dt} = \mathbf{i}\frac{dV_x}{dt} + \mathbf{j}\frac{dV_y}{dt} + \mathbf{k}\frac{dV_z}{dt}. \tag{7.27}$$

In such an expansion it is assumed that the unit vectors are constant with respect to time.

7.8.2 Spherical Polar Coordinates

In spherical polar coordinates x, y, and z are curvilinear distances and the unit vectors **i**, **j**, and **k**, which are tangential and normal, must change directions. Therefore, all terms involving derivatives of unit vectors must take this into account. (Unit vectors cannot change magnitude only direction as pointed out in section 2.6.3.) In figure 7.3 the change of each unit vector must be considered relative to distances west to east, south to north and up towards the zenith. Expansion of $d\mathbf{V}/dt$, where the unit vectors are now variable, gives

$$\frac{d\mathbf{V}}{dt} = \mathbf{i}\frac{du}{dt} + \mathbf{j}\frac{dv}{dt} + \mathbf{k}\frac{dw}{dt} + \frac{d\mathbf{i}}{dt}u + \frac{d\mathbf{j}}{dt}v + \frac{d\mathbf{k}}{dt}w. \tag{7.28}$$

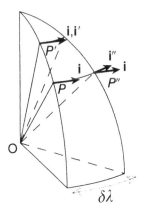

Figure 7.4 Movement of **i** from P to P'' and to P' respectively

Now each of the total derivatives of the unit vectors **n** may be expanded in term of partials (Equation 2.66),

$$\frac{d\mathbf{n}}{dt} = \frac{\partial \mathbf{n}}{\partial t} + \frac{\partial \mathbf{n}}{\partial x}u + \frac{\partial \mathbf{n}}{\partial y}v + \frac{\partial \mathbf{n}}{\partial z}w. \tag{7.29}$$

The unit vectors neither change with time locally nor with height, i.e. $\partial \mathbf{n}/\partial t = \partial \mathbf{n}/\partial z = 0$. Therefore, the first and last terms in Equation (7.29) vanish giving

$$\frac{d\mathbf{n}}{dt} = \frac{\partial \mathbf{n}}{\partial x}u + \frac{\partial \mathbf{n}}{\partial y}v. \tag{7.30}$$

Below, each unit vector $(\mathbf{i},\mathbf{j},\mathbf{k})$ in turn is substituted for **n** in Equation (7.30) and its components resolved. Then they will be combined into Equation (7.28).

Vector **i**

Figure 7.4 shows that there is no change in the **j** direction: **i** remains parallel to itself at each P' so $\partial \mathbf{i}/\partial y = 0$. Therefore, Equation (7.30) with substitution reduces to

$$\frac{d\mathbf{i}}{dt} = \frac{\partial \mathbf{i}}{\partial x}u. \tag{7.31}$$

In this direction, west–east, **i** is turned to $\mathbf{i}''$. As shown in figure 7.5(a) $\delta \mathbf{i}$ (assuming $\delta \lambda$ is small) points towards the axis of the earth. From figure 7.5(a) $\delta i = i\delta\lambda = \delta\lambda$, and $\delta x = r \cos \phi \delta\lambda$, so

$$\frac{\partial i}{\partial x} = \lim_{\delta x \to 0} \frac{\delta i}{\delta x} = \frac{\delta\lambda}{r \cos \phi \delta\lambda} = \frac{1}{r \cos \phi}.$$

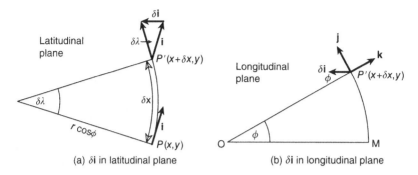

Figure 7.5 Two views of $\delta\mathbf{i}$, the difference between $\mathbf{i}''$ and $\mathbf{i}$, which is directed towards the earth's axis. Note that $\delta\lambda$ is small

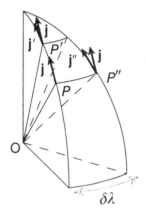

Figure 7.6 Movement of $\mathbf{j}$ from P to P'' and to P' respectively

But, as may be seen in figure 7.5(b), $\delta\mathbf{i}$ is directed towards the axis of the earth, so it has components in the $\mathbf{j}$ and $-\mathbf{k}$ directions. Thus

$$\frac{d\mathbf{i}}{dt} = \frac{\partial\mathbf{i}}{\partial x}u = \frac{u}{r\cos\phi}(\mathbf{j}\sin\phi - \mathbf{k}\cos\phi). \qquad (7.32)$$

Vector $\mathbf{j}$

The components for the change in the unit vector $\mathbf{j}$ are shown in figure 7.6. At P'' $\delta\mathbf{j}$ would be represented by the vector joining the heads of $\mathbf{j}$ and $\mathbf{j}''$ in figure 7.7(a). The angle between them is $\delta\theta$ which is defined by the triangle $P''NP'$ where N is on the extension of the north polar axis. Since $\delta j = j\,\delta\theta = \delta\theta$,

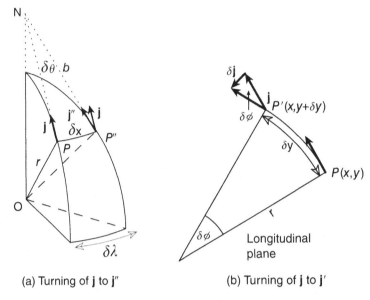

(a) Turning of **j** to **j**″   (b) Turning of **j** to **j**′

Figure 7.7 Turning of **j**. In (a) **j**″ makes an angle $\delta\theta$ with **j** as defined by the projection of these vectors to the extension of the north polar axis where they meet. In (b) $\delta\mathbf{j}$ is directed towards the center of the earth

$b = r/\tan\theta$ and $\delta x = b\,\delta\theta$

$$\frac{\partial j}{\partial x} = \lim_{\delta x \to 0} \frac{\delta j}{\delta x} = \frac{\delta x \tan\phi}{\delta x\, r} = \frac{\tan\phi}{r}.$$

This vector points in the $-\mathbf{i}$ direction so

$$\frac{\partial \mathbf{j}}{\partial x} = -\mathbf{i}\,\frac{\tan\phi}{r}. \tag{7.33}$$

$\delta\mathbf{j}$ at P' is shown in figure 7.7(b). Because $\delta j = j\,\delta\phi$ and $\delta y = r\,\delta\phi$

$$\frac{\partial j}{\partial y} = \lim_{\delta y \to 0} \frac{\delta i}{\delta y} = \frac{\delta\phi}{r\,\delta\phi} = \frac{1}{r}.$$

This vector points in the $-\mathbf{k}$ direction. Therefore

$$\frac{\partial \mathbf{j}}{\partial y} = -\mathbf{k}\,\frac{1}{r}. \tag{7.34}$$

Together Equations (7.33) and (7.34) give

$$\frac{d\mathbf{j}}{dt} = -\mathbf{i}\,\frac{u\tan\phi}{r} - \mathbf{k}\,\frac{v}{r}. \tag{7.35}$$

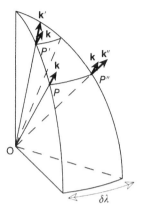

Figure 7.8 Turning of **k**. **k** to **k**′ is directed towards **j** and **k** to **k**″ is directed towards **i**

Vector **k**

The two resulting vectors are shown in figure 7.8. The $\partial \mathbf{k}/\partial x$ term is derived from the relationships $\delta k = k\,\delta\lambda = \delta\lambda$ and $\delta x = r\,\delta\lambda$ at P''. It points in the **i** direction so

$$\frac{\partial \mathbf{k}}{\partial x} = \mathbf{i}\,\frac{1}{r}.$$

Similarly at P'

$$\frac{\partial \mathbf{k}}{\partial y} = \mathbf{j}\,\frac{1}{r}$$

so

$$\frac{d\mathbf{k}}{dt} = \mathbf{i}\,\frac{u}{r} + \mathbf{j}\,\frac{v}{r}. \tag{7.36}$$

Substitution of Equations (7.32), (7.35), and (7.36) into Equation (7.28) with rearrangement of terms leads to

$$\frac{d\mathbf{V}}{dt} = \mathbf{i}\left(\frac{du}{dt} - \frac{uv\tan\phi}{r} + \frac{uw}{r}\right)$$
$$+ \mathbf{j}\left(\frac{dv}{dt} + \frac{u^2\tan\phi}{r} + \frac{vw}{r}\right) \tag{7.37}$$
$$+ \mathbf{k}\left(\frac{dw}{dt} - \frac{u^2 + v^2}{r}\right).$$

This is the full expansion of the local acceleration, $d\mathbf{V}/dt$, in spherical polar coordinates.

Complete Equations

The complete equations of motion in spherical polar coordinates are produced from Equations (7.26) and (7.37) with $dx = r\cos\phi\,d\lambda$ and $dy = r\,d\phi$:

$$\frac{du}{dt} - \frac{uv\tan\phi}{r} + \frac{uw}{r} = -\frac{1}{\rho\,r\cos\phi}\frac{\partial p}{\partial\lambda} + 2v\Omega\sin\phi - 2w\Omega\cos\phi + F_x,$$
(7.38)

$$\frac{dv}{dt} + \frac{u^2\tan\phi}{r} + \frac{vw}{r} = -\frac{1}{\rho\,r}\frac{\partial p}{\partial\phi} - 2u\Omega\sin\phi + F_y, \qquad (7.39)$$

$$\frac{dw}{dt} - \frac{u^2 + v^2}{r} = -\frac{1}{\rho}\frac{\partial p}{\partial z} - g + 2u\Omega\cos\phi + F_z. \qquad (7.40)$$

Vector Differential Operator in Spherical Polar Coordinates

Since the vector differential operator occurs in numerous equations, e.g. continuity, its expression in spherical polar coordinates is stated here.

$$\nabla = \mathbf{i}\,\frac{1}{r\cos\phi}\frac{\partial}{\partial\lambda} + \mathbf{j}\,\frac{1}{r}\frac{\partial}{\partial\phi} + \mathbf{k}\,\frac{\partial}{\partial z}. \qquad (7.41)$$

7.8.3 Natural Coordinates

A similar approach may be applied to produce natural coordinates in which the system is oriented so that one unit vector, $\mathbf{t}$, lies parallel to the horizontal flow, U (parallel to the horizontal streamline); one, $\mathbf{n}$, is perpendicular to U along the radius of curvature, R; and one, $\mathbf{k}$, is vertical (Haltiner and Martin, 1957),

$$\frac{d\mathbf{V}}{dt} = \mathbf{t}\,\frac{dU}{dt} + \mathbf{n}\,\frac{U^2}{R} - \mathbf{k}\,\frac{U}{r}, \qquad (7.42)$$

where r is the radius of the earth. The complete equations in this system are:

$$\frac{dU}{dt} = -\frac{1}{\rho}\frac{\partial p}{\partial s} + F_s, \qquad (7.43)$$

$$\frac{U^2}{R} = -\frac{1}{\rho}\frac{\partial p}{\partial n} - 2U\Omega\sin\phi + F_t, \qquad (7.44)$$

$$-\frac{U}{r} = -\frac{1}{\rho}\frac{\partial p}{\partial z} - g + 2u\Omega\cos\phi + F_z, \qquad (7.45)$$

where s and n are distances in the tangential (**t**) and normal (**n**) directions respectively.

7.8.4 Vertical Coordinate

Several different vertical coordinates have been developed and are in use.

Height, z, while appearing to be the natural vertical coordinate, produces relatively complicated equations and causes problems at the earth's surface where height levels disappear and reappear as they intersect relief. Even so, regardless of the vertical coordinate used in calculations, we usually like to convert back to height in summarizing our results.

In atmospheric analysis, geometric height is usually replaced by *geopotential height*. The geopotential at z, $\Phi(z)$, is the potential energy of a unit mass at height z, and is defined by the equation

$$\Phi(z) = \int_0^z g\,dz, \tag{7.46}$$

where g is variable (see section 4.3.3) and $z = 0$ is mean sea level.

Geopotential height is then defined by

$$Z = \frac{\Phi(z)}{g_0}, \tag{7.47}$$

where g_0 is the global average acceleration of gravity ($g_0 = 9.80665 \text{ m s}^{-2}$).

Pressure, p, is a fundamental characteristic of a gas whereas height is not. It simplifies some physical relationships and isobaric surfaces have long been the base for the drawing of upper air maps. The advantage here is that the variable, density, does not appear in the equations because the pressure gradient force on a horizontal surface is replaced by the height gradient on a constant pressure surface. Furthermore, in a vertical sounding the measurement of pressure is more easily made than height. The former needs a simple aneroid barometer whereas accurate determination of the latter requires more sophisticated equipment such as radar or GPS (global positioning system).

Potential temperature, θ, has advantages for the plotting of adiabatic motion. A potential temperature surface is known as an isentropic surface.

$$\theta = T\left(\frac{p_0}{p}\right)^{\kappa}, \tag{7.48}$$

(see section 5.7).

Sigma, σ, originally introduced by Phillips (1957), is the pressure normalized by the surface pressure. Thus sigma at the surface of the earth is always 1. This has special advantages in finite difference calculations in irregular terrain. Hence it has been used extensively in numerical modeling. Sigma is given by

$$\sigma = \frac{p}{p_{sfc}}, \text{ or } = \frac{p - p_T}{p_{sfc} - p_T} \qquad (7.49)$$

where p_{sfc} is the surface pressure, and p_T is the pressure of the top of the model domain. Sigma follows the terrain so sigma surfaces slope steeply where the terrain is steep.

Eta, η, a variation of the σ system which follows a step mountain terrain, was introduced by Mesinger (1984). It has advantages in calculating the pressure gradient force in models where the earth's surface slopes steeply. It is the coordinate selected for the main forecast model used in the US at the end of the twentieth century (Black, 1994).

$$\eta = \frac{(p - p_T)}{(p_{sfc} - p_T)} \left[\frac{p_{ref}(z_{sfc}) - p_T}{p_{ref}(z = 0) - p_T} \right]. \qquad (7.50)$$

As is evident, the first term is sigma. The second term modifies sigma so that the bottom domain is mean sea level. As a result, the eta surfaces are almost horizontal yet the coordinate remains a pressure based system. The terrain is stepped to match eta and the horizontal components of the wind are set to zero where eta changes.

7.9 Some Simple Solutions

One way of simplifying equations is to estimate the typical magnitudes of the terms and to drop those that are small relative to others. This approach was first introduced into meteorology by Jules Charney (1948). That paper should be consulted to follow how he arrived at his conclusions concerning the validity of his final equations and their filtering properties. While the spectrum of motion is continuous, atmospheric energy is concentrated in certain scales so it is legitimate to isolate certain groups of phenomena. The approximated magnitudes are known as *characteristic* quantities. For *large* scale motion in the atmosphere associated with systems such as extra-tropical cyclones and anti-cyclones, the horizontal extent is in the order of 1,000 km over which pressure changes by about 10 mb; the vertical extent is the height of the troposphere, 10 km; the horizontal winds, 10 m s^{-1}; the vertical winds, 10^{-2} m s^{-1}; and their time extent is several days, although for consistency it is given as horizontal length divided by speed. Table 7.2 provides a summary. F is estimated for the free atmosphere (say, one kilometer above the surface).

Table 7.2 Characteristic magnitudes for synoptic scales

L	10^6	m	r	10^6	m
H	10^4	m	g	10	$\mathrm{m\,s^{-2}}$
u, v	10	$\mathrm{m\,s^{-1}}$	Ω	10^{-4}	$\mathrm{s^{-1}}$
w	10^{-2}	$\mathrm{m\,s^{-1}}$	ρ	1	$\mathrm{kg\,m^{-3}}$
∇p	10^3	Pa	$F_{\phi,\lambda}$	10^{-4}	$\mathrm{m\,s^{-2}}$
$t = L/H$	10^5	s	F_z	10^{-5}	$\mathrm{m\,s^{-2}}$

7.9.1 Synoptic Scale Approximations

If we write down the typical magnitudes that we observe on synoptic maps for each of the terms in Equations (7.38)–(7.40) we get respectively

$$10^{-4} \quad 10^{-4} \quad 10^{-6} = 10^{-3} \quad 10^{-3} \quad 10^{-6} \quad 10^{-4}$$
$$10^{-4} \quad 10^{-4} \quad 10^{-6} = 10^{-3} \quad 10^{-3} \quad 10^{-4}$$
$$10^{-7} \quad 10^{-2} = 10 \quad\;\; 10 \quad\;\; 10^{-3} \quad 10^{-5}.$$

Now, if we drop all the smaller terms, we end up with the approximate results

$$0 = -\frac{1}{\rho}\frac{\partial p}{r\cos\phi\,\partial\lambda} + 2v\Omega\sin\phi, \tag{7.51}$$

$$0 = -\frac{1}{\rho}\frac{\partial p}{r\,\partial\phi} - 2u\Omega\sin\phi, \tag{7.52}$$

$$0 = -\frac{1}{\rho}\frac{\partial p}{\partial z} - g. \tag{7.53}$$

These are simply diagnostic equations describing the approximate balances of forces in the atmosphere. Their significance and application are discussed below.

The Geostrophic Wind

Equations (7.51) and (7.52) display a balance between the horizontal pressure gradient and the Coriolis force. The term $2\Omega\sin\phi$, which is known as the *Coriolis parameter*, is usually replaced by f. These equations may be solved for the wind, giving the *geostrophic* wind components u_g and v_g. They describe the theoretical winds that blow parallel to the isobars (or contour lines on a constant pressure map) with low on the left in the northern hemisphere and on the right in the southern (see figure 7.9 for two examples).

The geostrophic wind does not imply any cause and effect relationship. It states that the pressure gradient and the geostrophic wind velocity are a pair of elements each of which is specified by its partner. This makes the relationship extremely useful. It means that one can replace the other. Given a pressure map the wind field is immediately available. It should be noted that the speed of the geostrophic wind is also given by the pressure gradient, i.e. the steeper the gradient, the closer the isobars, the greater the speed. Also the speed varies with the Coriolis parameter. Therefore, with a constant gradient the speed increases towards lower latitudes. However, at the Equator $f = 0$ so the geostrophic wind becomes infinity and becomes meaningless.

The accuracy of the geostrophic approximation depends upon the relative magnitudes of the terms in Equations (7.38) and (7.39). At the synoptic scale above the friction layer the curvature, for example, may affect the wind significantly. In such cases the *gradient wind* approximation provides a more accurate representation. For the same gradient, cyclonically curved flow has a smaller gradient speed than the geostrophic speed. [For a fuller discussion, see for example, Holton (1979) section 3.2.5.]

As a measure of the importance of acceleration the ratio of the inertial force to the acceleration force is used,

$$R_o = \frac{U^2/L}{fU} = \frac{U}{fL}. \tag{7.54}$$

This is now called the *Rossby number*. The smaller the number, the more accurate the geostrophic wind.

Godson (1950) compared geostrophic and observed winds and concluded:

(a) For latitudes north of $30°$ and for geostrophic wind speeds exceeding $20\,\mathrm{mi\,hr^{-1}}$ the angle between the observed wind and the geostrophic wind is of the order of $10°$, considerably larger values being possible if these conditions are not fulfilled.

(b) On the average the cross-contour wind component is of the order of 20 per cent of the wind speed. The "isallobaric" contribution to this component is generally smaller than the contributions ascribable to horizontal or vertical "velocity advection."

(c) For latitudes north of $30°$ and for geostrophic wind speeds between 20 and $45\,\mathrm{mi\,hr^{-1}}$ the difference in speed between observed and geostrophic winds is of the order of 20 per cent of the true wind speed, considerably larger values being possible if these conditions are not fulfilled.

(d) For latitudes south of $30°$ and/or for low geostrophic wind speeds the geostrophic wind equation gives a poor approximation to the actual wind; no significant improvement is possible by any existing methods of computing gradient winds.

(e) For latitudes north of $30°$ and for geostrophic wind speeds from 13 to $30\,\mathrm{mi\,hr^{-1}}$ the geostrophic wind equation gives a fair approximation to the actual wind; slight improvement may be possible by consideration of contour curvature and the motion of contour systems.

(f) For geostrophic wind speeds exceeding $30\,mi\,hr^{-1}$ the geostrophic wind equation gives a poor approximation to the actual wind; at middle and high latitudes a better approximation would be afforded by the gradient wind equation if the trajectory curvature could be evaluated with sufficient accuracy.

The deviation of the actual wind from the geostrophic is known as the *a-geostrophic* component.

Thermal Wind

The difference in geopotential between two levels in a hydrostatic atmosphere, known as *thickness*, can be shown to be proportional to the average temperature of the layer. Since the geostrophic wind can be expressed in terms of the the geopotential field, the difference between the geostrophic wind at the top and the bottom of a layer can be related to thickness field and, therefore the temperature. The geostrophic wind difference, obtained graphically by joining the head of the lower vector to the head of the upper vector, is called the *thermal* wind (see figure 7.9). It is a theoretical construct but it stands in a similar relation to the temperature field as the geostrophic wind relates to the height field, i.e. low (height, temperature) is on the left of the (geostrophic, thermal) wind in the northern hemisphere.

The Hydrostatic Equation

Equation (7.53) shows that in the large scale the weight of the atmosphere is balanced by the vertical pressure gradient. The relationship is known as the *hydrostatic equation*.

Discussion

Not only do the geostrophic and hydrostatic approximations simplify the relationships between forces as shown, they are valuable in filtering out unwanted fluctuations, such as sound and gravity waves, that affect atmospheric calculations (Charney, 1948). Consequently, although these equations cannot be used directly for prediction, they may be used judiciously in simulation models for their filtering properties.

At scales smaller than the synoptic the characteristic quantities change. Thus numerical models now usually employ non-hydrostatic assumptions.

At the small scale, it is the Coriolis force that is frequently ignored. Many discussions of the sea breeze and mountain and valley winds adopt a non-Coriolis approach although a number of texts emphasize its importance (Atkinson, 1981; Simpson, 1994). Even at the scale of everyday human experience we may not be able to dismiss the deflection effect of earth rotation

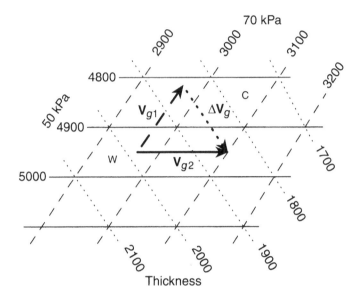

Figure 7.9 The geostrophic and thermal wind in the northern hemisphere. V_{g1} and V_{g2} are the geostrophic winds at the bottom and top of the 70 and 50kPa layer respectively. ΔV_g is the thermal wind. In this example the geostrophic winds throughout the layer will advect warm air

(Persson,1998). As is illustrated in one of the early exercises in Holton (1979) a typical baseball pitch will change direction by more than a centimeter.

The conclusion to be drawn is that, while simplifications are valuable especially for generalizing arguments about atmospheric flow, there is a limit to their application. They must be employed with care. Unfortunately, because these relationships usually fit observed data very well many users forget this fact.

7.10 Fluid Rotation

Just as rotation is an important characteristic of solid bodies, it is very significant in fluids. Now, however, its specification is more complicated because each element of the fluid can move independently and the angular velocities on different axes are not necessarily related. If we apply the definition of angular velocity from Equation (4.7) as shown in figure 7.10 where the center of the rotation is at the center of the coordinate system, we get two velocities:

$$\omega_1 = \frac{v}{x}, \qquad \omega_2 = -\frac{u}{y}.$$

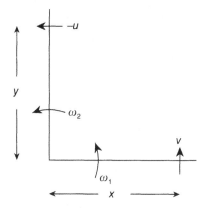

Figure 7.10 Angular velocity on two axes

But as pointed out earlier, Equation (4.35) expresses the angular velocity for an element of fluid without regard to the center of rotation. Applied to the xy plane of figure 7.10 we obtain

$$\frac{\partial v}{\partial x} = \omega_1,$$

and

$$-\frac{\partial u}{\partial y} = \omega_2.$$

The total angular velocity in this plane is the sum

$$\zeta = \frac{\partial v}{\partial x} - \frac{\partial u}{\partial y} = \omega_1 + \omega_2. \tag{7.55}$$

Since, for a solid body $\omega_1 = \omega_2$, Equation (7.55) expresses twice this velocity. In fact, von Helmholtz (1858) put a half in the equation defining ζ. Lamb (1932) was apparently the first to drop the half. The rotation velocities in the other planes are

$$\xi = \frac{\partial w}{\partial y} - \frac{\partial v}{\partial z}, \tag{7.56}$$

and

$$\eta = \frac{\partial u}{\partial z} - \frac{\partial w}{\partial x}. \tag{7.57}$$

In vector form these become

$$\nabla \times \mathbf{V} = \xi \mathbf{i} + \eta \mathbf{j} + \zeta \mathbf{k}. \tag{7.58}$$

As identified in section 2.6.4 this is also called the curl of the vector. In spherical polar coordinates ζ becomes

$$\zeta = \frac{1}{R \cos \phi} \frac{\partial v}{\partial x} - \frac{1}{R} \frac{\partial u}{\partial \phi} + \frac{u \tan \phi}{R}. \tag{7.59}$$

We may use a different approach to arriving at these equations through considering the circulation around an element of fluid, that is, around a vortex (Lamb, 1932, 33–5, 247). Consequently, these equations are said to describe the *vorticity* of the fluid.

Although circulation was investigated early by von Helmholtz and by Kelvin under restricted conditions (Haurwitz, 1941, 134–44) its application to the atmosphere (and oceans) is due to Bjerknes (1898). Vilhelm Bjerknes (1862–1951) was born in Oslo. As a professor of mechanics and mathematical physics at the University of Stockholm, he initially continued his father's research into long range magnetic and electric forces but his methods, which included the ether, were not considered mainstream physics for the turn of the century. However, it was in applying his force field concepts to hydrodynamics that he developed his circulation theorem. In 1905 he was invited to present lectures on physics at Columbia University and, while in the US, he visited Washington where he met and impressed Cleveland Abbe. As a result he received a Carnegie Institution Grant in 1906 that led to the publication of *Dynamic Meteorology and Hydrography* in 1910. During his stay in Stockholm he became great friends with Svante Arrhenius, the chemist who wrote on the greenhouse effect. Partly due to the breakup of the union between Sweden and Norway in 1905 Bjerknes moved to the University of Christiania in 1907. Unfortunately, Bjerknes' ideas applied to meteorology were seen somewhat as revolutionary in established quarters in Europe and he was unable to secure the support for obtaining the observations he needed for his theories (Friedman, 1989). In 1912 he was invited to set up a geophysical institute in Leipzig. There, even though he remained neutral and lost many of his scientists to World War I, he continued his meteorological research. In 1917 he returned to Norway taking two students, Jacob Bjerknes, his son, and Halvor Solberg, with him. He convinced the Norwegian government, which was dealing with serious food shortages, to improve the Norwegian Meteorological Service to serve agriculture and fisheries. The outcome was the establishment of the famous Bergen School and an increase in meteorological stations in Southern Norway from 9 to 90 (Norske-Videnskaps, 1962). In 1926 Vilhelm was appointed to a professorship at the University of Oslo where he remained even after he retired in 1932.

Usually in atmospheric work only the vertical component, ζ, of vorticity is used. A view of what that component of vorticity represents may be gained from considering it in natural coordinates,

$$\zeta = \frac{V}{R} - \frac{\partial V}{\partial n}. \tag{7.60}$$

This states that vorticity is positive for counter-clockwise rotation in the northern hemisphere (cyclonic) and positive for shear where the wind speed increases from low to high pressure. As an example, in a cyclone where the tangential wind increases towards the center the two components may offset one another such that the total local vorticity may become zero or even negative.

So far we have considered the relative wind, V_{rel}. If we substitute the absolute velocity into the curl we obtain

$$\mathbf{k} \cdot \nabla \times \mathbf{V}_{inertial} = \mathbf{k} \cdot (\nabla \times \mathbf{V}_{rel} + 2\Omega)$$
$$= \zeta + f. \tag{7.61}$$

This states that the total, or absolute, vorticity is the sum of the local, or relative (ζ), vorticity and the earth's vorticity (f).

As emphasized in the discussion of angular momentum (section 4.3.7) rotation is a conservative quantity and turns out to be very useful in characterizing fluid flows. Indeed, from the above discussion it can be seen that vorticity is to be expected whenever friction exists to produce shear. Therefore, we find vorticity and its visible expression in wave and eddy motion present throughout the atmosphere and at various scales.

The factors that change absolute vorticity are obtained from its time derivative. Various expressions of this equation are available depending upon the selection of the coordinate system and vertical coordinate. In Cartesian coordinates it is:

$$\frac{d(\zeta + f)}{dt} = -(\zeta + f)\left(\frac{\partial u}{\partial x} + \frac{\partial v}{\partial y}\right)$$
$$+ \left(\frac{\partial w}{\partial y}\frac{\partial u}{\partial z} - \frac{\partial w}{\partial x}\frac{\partial v}{\partial z}\right)$$
$$+ \frac{1}{\alpha}\left(\frac{\partial \alpha}{\partial y}\frac{\partial p}{\partial x} - \frac{\partial \alpha}{\partial x}\frac{\partial p}{\partial y}\right), \tag{7.62}$$

where $\alpha = 1/\rho$. The resulting factors are 1) convergence (effect of an ice skater withdrawing spread out arms to increase spin); 2) conversion from the other two components through the tilting of a vortex; and 3) the pressure torque, called the *solenoidal* component. This term

$$\left(\frac{\partial \alpha}{\partial y} \frac{\partial p}{\partial x} - \frac{\partial \alpha}{\partial x} \frac{\partial p}{\partial y} \right),$$

is what defines *baroclinicity*, the degree to which the isobaric and isopycnic surfaces intersect. Parallel surfaces indicate a *barotropic* atmosphere.

The solenoidal term vanishes from the vorticity equation if the vertical coordinate is pressure. In addition, the third factor vanishes for dry adiabatic motion if the vertical coordinate is potential temperature. Through applying simplifying assumptions Rossby was able to develop his well-known wave equation (Rossby and Collaborators, 1939).

Carl-Gustaf Rossby (1898–1957) was one of the most influential meteorologists of the twentieth century. Born in Stockholm he was interested in a broad range of fields and only chose meteorology late in his education. He completed his undergraduate degree in astronomy, mathematics, and mechanics. In 1919 at the recommendation of one of his mathematics professors he went to work for V. Bjerknes in Bergen where he was an associate of Bjerknes' son Jacob, Bergeron, and Solberg. Later, while at the Swedish Meteorological and Hydrological Institute he completed a doctorate at Stockholm in 1925 (Bergeron, 1959). That year he moved to the United States to work for the Weather Bureau where his ideas were hardly tolerated (Byers, 1959). In 1927 he was appointed chairman of the Guggenheim Fund's Committee on Aeronautical Meteorology in which capacity he set up an experimental weather service for an airline between San Francisco and Los Angeles. A year later he was appointed to MIT where a new department of meteorology was established with Rossby as the first chair, in fact the first chair of meteorology in the US (Sutcliffe, 1958). In 1939 Rossby again took a position in the Weather Bureau where reaction toward him improved with the appointment of Reichelderfer as Chief. Together they established a training program around the country to upgrade the weather personnel. Rossby's interest in training, especially for the military, continued after he moved to the University of Chicago in 1941. Many hundreds of individuals received intensified courses under that program including several geographers. With Palmer he also established a Tropical Institute in Puerto Rico where Riehl did his early work on tropical meteorology (Byers, 1959). In 1947 he returned to Scandinavia to accept a special chair at the University of Stockholm where he died in 1957 (Bolin, 1999).

Rossby's most important direct research contribution to meteorology was his work on vorticity and long waves but his organizational and inspirational contributions were perhaps more significant.

7.11 The Equation Set

The basic equations describing the atmosphere are seven in number. For clarity we shall repeat them here.

Continuity – Equation (7.4)

$$\frac{1}{\rho}\frac{d\rho}{dt} + \nabla \cdot \mathbf{V} = 0 \tag{7.63}$$

Equation of motion – Equation (7.26)

$$\frac{d\mathbf{V}}{dt} = -(1/\rho)\nabla p - (2\mathbf{\Omega} \times \mathbf{V}) - \mathbf{g} + \mathbf{F}. \tag{7.64}$$

This actually produces three equations, one for each component direction.

Equation of State – Equation (5.51)

$$p = \rho R_d(1 + 0.608q)T. \tag{7.65}$$

Thermodynamic Equation – Equation (5.27)

$$Q = c_p \frac{dT}{dt} - \alpha \frac{dp}{dt}. \tag{7.66}$$

The Conservation of Water – Equation (7.67)

$$\frac{dq}{dt} = B, \tag{7.67}$$

where B represents flow through the boundaries of a parcel (for example, precipitation, evapotranspiration, and molecular diffusion).

The Set

The above equations may appear in different forms. Sometimes the vorticity is used. However, we still return to seven equations composed of seven variables: T, ρ, p, q, u, v, and w; specified in four dimensional space: λ, ϕ, z, and t. For a complete set, other variables, such as CO_2 and its associated equations must be added.

Also, a number of terms involve undefined processes. Friction, **F**, is large near the surface, and vertical turbulent transfers tend to be larger than horizontal ones but they must all be incorporated in an explicit manner. Diabatic heat flow, Q, involves turbulent transfers as well as radiation and latent heat conversions. Each of these processes takes place on a small scale and typically they must be integrated over larger volumes before they can enter the above equations. The means by which this is done is referred to as *parameterization.*

7.12 Comment

The material in this chapter is only a brief introduction to the fluid dynamics of the atmosphere. Lamb's *Hydrodynamics* is still an excellent reference for this field although numerous texts, written specifically for the atmosphere, have appeared. Holton (1979) is still currently widely used but there are many others typically having the word *dynamic* in the title (Dutton (1976) is an exception). Most authors manipulate the equations so as to explain various characteristic flows and systems. These range from local mountain waves to the large scale circulation. They are partially successful, being limited in their generality by the assumptions imposed. Future research in combination with numerical simulation is expected to enhance this approach.

Chapter 8

Observed Angular Momentum and Energy

8.1 Perspective

Typically global climate is represented by maps, cross-sections, and tables of atmospheric elements such as pressure, winds, temperature, precipitation, etc. They may be found in hundreds of introductory climate textbooks. While these displays are basic components of climate, it has been argued in chapter 3 that they are usually inadequate descriptions of such data. Moreover, these are only limited expressions of the full climatic system: that is, all the processes that result from incoming solar radiation interacting with the earth and atmosphere.

One approach to synthesizing this system is to consider the overarching concept of entropy whereby energy is naturally dispersed from an ordered state into an unordered state. The concept was originally developed in thermodynamics (see section 5.8) where, in the simplest examples, heat could not spontaneously flow from a cold object to a warm object or be converted into mechanical energy. It dictates the direction of flow. Thus, dispersed solar energy is unevenly intercepted by the earth where it is dispersed within that system and eventually dispersed to space. So, temperature gradients are established and maintained against the processes that tend to weaken and destroy them.

Energy then is the central commodity. Every process involves energy, a conversion and a flow. Furthermore, we find that energy is indestructible. It can be neither created nor destroyed. It is conserved. Even though energy and mass are not considered synonymous in climate, the concept of conservation applies equally to mass. In addition, because the earth–atmosphere is a rotating system, its angular momentum is also an important property that needs to be incorporated. An obvious way to describe the conversions and flows of energy, mass, and momentum is to run an accounting system on them. As a result the literature on climate is full of research on "Budgets of . . . " and "Balances of . . . ." Usually these deal only with subsets, such as "heat" or "water," but they are clearly part of the whole system.

In order to gain a perspective on the relative magnitudes of energy Sellers (1965) provided two tables, extracts of which are given in tables 8.1 and 8.2.

The relative cost of some forms of energy in the United States are given in table 8.3.

8.2 Angular Momentum

8.2.1 Total Angular Momentum

Through rotation, the earth and atmosphere possess angular momentum. If the earth had uniform density, Equation (4.36) gives an estimate of

$$
\begin{aligned}
I_e \omega &= \frac{2}{5} M_e r^2 \Omega \\
&= \frac{2}{5} \times 5.975 \times 10^{24} \times (6.371 \times 10^6)^2 \times 7.292 \times 10^{-5} \quad (8.1) \\
&= 7.07 \times 10^{33} \, \text{kg} \, \text{m}^2 \, \text{s}^{-1}.
\end{aligned}
$$

Because density decreases towards the surface this is an over-estimate and Smith and Dahlen (1981) calculate the real moment of inertia to be 8.0438 kg m² making the earth momentum to be closer to 6×10^{33} kg m² s⁻¹.

For the atmosphere, assuming it to be in solid rotation with the earth, we can calculate its angular momentum from Equation (4.38),

$$
\begin{aligned}
\frac{2}{3} M_{sh} r^2 \omega &= \frac{2}{3} \times 5.15 \times 10^{18} \times (6.371 \times 10^6)^2 \times 7.292 \times 10^{-5} \quad (8.2) \\
&= 1.02 \times 10^{28} \, \text{kg} \, \text{m}^2 \, \text{s}^{-1}.
\end{aligned}
$$

This means that the angular momentum of the earth is about 10^6 times that of the atmosphere.

8.2.2 Angular Momentum Transfer

Momentum is transmitted between the earth and the atmosphere where their relative zonal speeds differ. Surface west wind components appear where the atmospheric shell is traveling more quickly than the earth and east wind components where it is traveling more slowly. The latter generally occurs equatorward of the sub-tropical highs and the former poleward of the same highs. Because the earth and atmosphere both have overall westerly momentum,

Table 8.1 Total energy of various individual phenomena and localized processes in the atmosphere. Rates are relative to total solar energy intercepted by the earth (1.5×10^{22} J day^{-1} or 1.75×10^{17} W)

Solar energy received per day	1
Melting of average winter snow during the spring season	10^{-1}
Monsoon circulation	10^{-2}
World use of energy in 1950	10^{-2}
Strong earthquake	10^{-2}
Average cyclone	10^{-3}
Average hurricane	10^{-4}
Krakatoa explosion of August, 1883	10^{-5}
Detonation of "thermonuclear weapon" in April, 1954	10^{-5}
Kinetic energy of the general circulation	10^{-5}
Average squall line	10^{-6}
Average summer thunderstorm	10^{-8}
Detonation of Nagasaki bomb in August, 1945	10^{-8}
Average earthquake	10^{-8}
Average local shower	10^{-10}
Average tornado	10^{-11}
Street lighting on average night in New York City	10^{-11}
Average Lightning stroke	10^{-13}
Average dust devil	10^{-15}
Individual gust near the earth's surface	10^{-17}

Table 8.2 Large-scale energy sources. Rates are relative to solar energy available (11×10^9 J m^{-2} yr^{-1} or 3,492 Wm^{-2}). Sellers used a solar constant of 1,396 Wm^{-2} which may be compared to a currently accepted magnitude of 1,376 Wm^{-2}

One-fourth solar constant (see Equation 6.32)	1
Heat flux from the earth's interior	18×10^{-5}
Infrared radiation from the full moon	3×10^{-5}
Sun's radiation reflected from the full moon	1×10^{-5}
Energy generated by solar tidal forces in the atmosphere	1×10^{-5}
Combustion of coal, oil, and gas in the United States	7×10^{-6}
Energy dissipated in lightning discharges	6×10^{-7}
Total radiation from stars	4×10^{-8}
Energy generated by lunar tidal forces in the atmosphere	3×10^{-8}

the sign of that momentum is, by convention, considered positive. Therefore, westerly momentum is generally transferred from the earth to the atmosphere in low latitudes, and from the atmosphere to the earth in middle and high latitudes (except where there are polar easterlies).

Table 8.3 Cost of energy in the United States with 1999 values

Electricity	$0.07\,\$\,(\mathrm{kW\,hr})^{-1}$	$0.02\,\$\,\mathrm{MJ}^{-1}$
Natural gas	$0.70\,\$\,(100\,\mathrm{ft}^3)^{-1}$	$0.01\,\$\,\mathrm{MJ}^{-1}$
Gasoline	$1.30\,\$\,(\mathrm{US\,gal})^{-1}$	$0.01\,\$\,\mathrm{MJ}^{-1}$
Sugar	$1.75\,\$\,(5\,\mathrm{lb})^{-1}$	$0.05\,\$\,\mathrm{MJ}^{-1}$

The equation for the angular momentum per unit mass is:

$$\mathcal{M} = \Omega\,(r+z)^2 \cos^2\phi + u(r+z)\cos\phi. \tag{8.3}$$

Only two torques may change this, the pressure gradient force and friction torques, i.e.

$$\frac{d\mathcal{M}}{dt} = \left(-\frac{1}{\rho}\frac{\partial p}{\partial x} + F_x\right) r. \tag{8.4}$$

If we rewrite this in terms of $\rho\mathcal{M}$, expand, and then integrate over a polar cap, we obtain (see, for example, Haltiner and Martin (1957)),

$$\frac{\partial}{\partial t}\int \rho\mathcal{M}\,\delta\mathcal{V} = \mathcal{T} + \mathcal{P} + \mathcal{F}. \tag{8.5}$$

$\mathcal{T}$ is the net momentum transport through the latitudinal wall. $\mathcal{P}$ is the total pressure torque around latitude circles. If a whole latitude circle is at sea level there is no $\mathcal{P}$ torque because the pressure is continuous and returns to its original magnitude. In contrast, when mountains are present, pressure is not continuous and pressure differences can occur across them. Hence $\mathcal{P}$ is the mountain torque. The final term $\mathcal{F}$ is the frictional torque. Since the earth and atmosphere maintain essentially a constant angular momentum the flows must be balanced. In other words, the atmosphere gains westerly momentum in low latitudes and loses it in high latitudes at the surface through the effect of $\mathcal{P}$ and $\mathcal{F}$. The difference is accounted for by the transfer poleward in the atmosphere.

It was Jeffreys (1926) who first showed that such a transfer was accomplished by a correlation between the u and v wind components, which were observable in the asymmetric mid-latitude cyclone model of J. Bjerknes (1919)

At this stage it becomes clear that $\rho u v$ has an intimate relation to the Reynolds shearing stress $\rho u'v'$ ... pressure anomalies in temperate region cyclones extend up to the tropopause, but the constitution of the cyclone as fundamentally a combination of south west and north east winds appears to agree with the model of Bjerknes rather than with the symmetrical model of earlier writers. (Jeffreys, 1926)

Because Jeffreys defined v positive from the north, his discussion has to be read carefully. The transport to which he referred may be written as:

$$\mathcal{T} = 2\pi (r + z)^2 \cos^2 \phi \int\limits_{z=0}^{\infty} (\Omega \overline{v}(r + z) \cos \phi + \overline{u}\,\overline{v} + \overline{u'v'}) \rho \, dz, \qquad (8.6)$$

where the overbar represents an average around the latitude and the prime is the variation from that. There can be no net transfer of mass so $\rho \overline{v}$ summed over height must be zero. Hence, since Ω is constant, the first term on the right, the omega term, can produce no net flow, only flow at one level offset by the reverse flow at another level (mean overturning). The second term is positive if both $\overline{u}$ and $\overline{v}$ increase with height. It is known as the drift term. The last term is due to the waves/eddies around the latitude circle.

Sir Harold Jeffreys (1891–1989) was the son of a schoolmaster in northern England. He completed his education at Cambridge where he stayed his whole life except for a few years (1917–22) in the Meteorological Office with Napier Shaw. He worked mainly in statistics and geophysics. According to his obituary,

> It has been said in jest that after joining the Meteorological Office in 1917 and being asked to work on dynamical meteorology, Harold Jeffreys considered the relative importance of all the terms in the governing mathematical equations, decided that none could be safely neglected and resigned! (Hide, 1989)

A more philosophical statement made by Jeffreys concerning conflicting views held by some of his contemporaries was:

> I think the source of the trouble is the belief that there is some special virtue in mathematics. Instead of being regarded for what it is, a tool for dealing with arguments too complicated to be presented without it, it has become emotionalised to such an extent that many people think that nothing but mathematics has any meaning . . . . Its function in science is to connect postulates with observations. The utility of deduction is that in investigating the consequences of a well-supported law it is a convenient approximation to induction. (Hide, 1989)

8.2.3 Estimates

The earliest estimates of momentum transfer, like those for the other fluxes, had to await sufficient measurements of variables in the free atmosphere. As a result of the demands of weather forecasting, the number of aircraft taking to the skies, and World War II, a world-wide network of radiosonde stations was established in the late 1930s and early 1940s. By 1941 Rossby was able to draw pressure maps at 3 km, "as determined with the aid of upper-air

data now available daily from a number of stations in the United States."
(Rossby, 1941)

The first estimates using data from one station were made by Priestley in the late 1940s. Since he had worked on the vertical transfer of heat near the ground using the Reynolds' resolution (Priestley and Swinbank, 1947) it appears to have been a simple step for Priestley to apply Jeffreys' ideas to observational data. He used the years 1946 and 1947 from Larkhill in southern England to calculate $\overline{u'v'}$, which he called the eddy-flux, and $\overline{u}\,\overline{v}$, which he called the toroidal flux. In this case the overbar represents the time mean. The eddy fluxes tended to increase from the surface to 250 mb (jet stream level) and decrease thereafter. At the level of maximum it varied from -104 to $114\,\mathrm{m^2\,s^{-2}}$. The toroidal flux displayed a northward flow in the lower layers below 450 mb and southward flow above, suggesting Ferrel Cell overturning (Priestley, 1949).

C. H. B. (Bill) Priestley (1916–98) graduated from Cambridge in applied mathematics in 1937, after which he joined the British Meteorological Office where he was part of the forecast team for the World War II invasion on D-day. He is recognized both for his very successful administrative career as the first Chief of the CSIRO Division of Meteorological Physics in Australia (1946–71) and his own research work into turbulent processes at all scales (Holper, 1999).

As data became more plentiful, broader analyses were performed not just for momentum but for heat, moisture, and energy. Much of this research was supported by the the Geophysics Research Division, Air Force Cambridge Research Center. For example, Mintz (1951) of UCLA calculated the geostrophic poleward flux of momentum for January 1949. In a related work Mintz and Dean (1952) documented the observed motion of the atmosphere. This material was incorporated into the geography text of Trewartha (1954) and made it the most advanced climatology book in the field. At about the same time researchers, Victor P. Starr at MIT and Robert M. White, who was then at the nearby Air Force Cambridge Research Center, initiated a similar study (Starr and White, 1954). For half a century the investigation has occupied large numbers of meteorologists, many of whom were associated with MIT. The most recent global estimates in this series were summarized by Peixoto and Oort (1992), both of whom studied at MIT. In an early paper Starr and White (1952) reviewed the formulae used for calculating the fluxes and how the order in which averaging was performed, latitude, longitude, height, and time, affected the interpretation of the Reynolds' terms. In their first northern hemispheric analysis for the year 1950 they confirmed Jeffreys' suggestion. They stated:

It is found that the flux of such momentum between tropical and middle latitudes is principally accomplished through the agency of large scale eddies in the atmosphere. (Starr and White, 1954)

Victor Paul Starr (1909–76) graduated from New York State College in Albany with a BS in chemistry in 1930, after which he worked in the US Weather Bureau. His research impressed Harry Wexler in the Bureau and he was encouraged to continue with his education. In 1937 he completed an MS at MIT where he worked with Rossby. Later Rossby invited him to Chicago for doctoral work that he completed in 1946. As a Professor of Meteorology at MIT from 1947 until he retired in 1974, Starr influenced several generations of meteorologists (Lorenz, 1977).

The relationship between the angular momentum fluxes and the shape of the waves and eddies as predicted by Jeffreys is clear (Starr, 1948). For momentum to be transferred the systems must be asymmetric. Similarly, the flux of other elements such as sensible and latent heat require phase differences between the wind and those variables. The northern and southern hemispheres present different viscosities and different mountain torques. Consequently, their circulations must be different. By the same argument different configurations of the continents in the geological past should have led to different circulation patterns. Also, despite the relatively small momentum of the atmosphere compared to the earth, the stresses do affect the length of day and may influence plate tectonics (Peixoto and Oort, 1992).

8.3 The Partition of Energy

Atmospheric energy is found in four forms: internal, kinetic, latent, and potential. These have been defined in the previous chapters by Equations (5.22), (4.25), (5.41), and (4.24) respectively. Restated in terms of unit mass they are

$$\text{Internal Energy (IE)} = c_v\, T, \qquad (8.7)$$

$$\text{Kinetic Energy (KE)} = \frac{1}{2} V^2, \qquad (8.8)$$

$$\text{Latent Energy (LE)} = l q, \qquad (8.9)$$

$$\text{Potential Energy (PE)} = g z. \qquad (8.10)$$

If climate is stable, then the average magnitudes of these energies, and consequently their sum, must remain constant. On the other hand, since the system is dynamic, energy is continually flowing through it. Therefore, over the period of the stable climate, these flows must be balanced.

The equations for the flows of energy may be obtained from taking the time derivatives of the energies in Equations (8.7) to (8.10).

Internal Energy

We have an equation for the time rate of change of internal energy from the First Law of Thermodynamics, Equation (5.26),

$$\frac{d(IE)}{dt} = c_v \frac{dT}{dt} = \mathcal{Q} - p \frac{d\alpha}{dt}.$$

If we substitute from Equation (7.4) this is

$$\frac{d(IE)}{dt} = \mathcal{Q} - \alpha p \nabla \cdot \mathbf{V}. \qquad (8.11)$$

Kinetic Energy

We can obtain the time rate of change of kinetic energy by dot multiplying the equation of motion (7.26) by $\mathbf{V}$,

$$\frac{d\mathbf{V}}{dt} \cdot \mathbf{V} = -\alpha \nabla p \cdot \mathbf{V} - (2\mathbf{\Omega} \times \mathbf{V}) \cdot \mathbf{V} - \mathbf{g} \cdot \mathbf{V} + \mathbf{F} \cdot \mathbf{V},$$

because, from Equations (2.23) and (2.46),

$$d(V^2/2) = V\,dV$$

which means that

$$\frac{d\mathbf{V}}{dt} \cdot \mathbf{V} = \frac{d(V^2/2)}{dt}.$$

Also, if we recognize from Equation (2.66), that $(2\mathbf{\Omega} \times \mathbf{V}) \cdot \mathbf{V} = 0$, and that $\mathbf{g} \cdot \mathbf{V} = gw$, we have

$$\frac{d(KE)}{dt} = \frac{d(V^2/2)}{dt} = -\alpha \nabla p \cdot \mathbf{V} - gw + \mathbf{F} \cdot \mathbf{V}. \qquad (8.12)$$

Latent Energy

The time derivative of latent heat is

$$\frac{d(LE)}{dt} = \frac{d(lq)}{dt} = lB. \qquad (8.13)$$

Although not explicitly appearing in Equation (8.11) the conversion of latent heat is part of $\mathcal{Q}$.

Potential Energy

Potential energy change can be written from Equation (7.46)

$$\frac{d(PE)}{dt} = \frac{d\Phi}{dt} = gw. \tag{8.14}$$

Rate of Total Energy Change

A review of Equations (8.11), (8.12), (8.13), and (8.14) reveals that they involve similar terms showing that these energies are interrelated. If we observe that

$$-\alpha p\nabla \cdot \mathbf{V} - \alpha\nabla p \cdot \mathbf{V} = -\alpha\nabla \cdot p\mathbf{V},$$

their sum reduces to:

$$\frac{d(c_v T + V^2/2 + lq + gz)}{dt} = Q - \alpha\nabla \cdot p\mathbf{V} + \mathbf{F} \cdot \mathbf{V} + \frac{d(lq)}{dt} \tag{8.15}$$

8.4 The Lorenz Model of Energy Flow

In 1955 Lorenz proposed a scheme for global atmospheric energy flow. Edward N. Lorenz (1917–), born in Hartford, CT, studied mathematics at Dartmouth (BA 1938) and Harvard (MA 1940). He was a Teaching Fellow at Harvard before serving as a weather forecaster in the US Air Corps from 1942–46. Meanwhile he pursued graduate work in meteorology at MIT receiving his masters in 1943 and doctorate in 1948. He remained as a faculty member at MIT until he retired in 1987. Besides his seminal publications in global energy, he is responsible, as mentioned in chapter 1, for the initial work and concept of chaos (Gedzelman, 1994; Lorenz, 1993). This was an outgrowth of his ideas about the limits of weather forecasting and also led to his innovative investigations concerning climatic and transitivity (see section 9.1.1). His fundamental contributions have been acknowledged through numerous honorary degrees and national and international awards. He was elected to the National Academy of Sciences in 1975.

In his paper, Lorenz (1955) essentially followed the arguments of Margules (1901, 1905) who showed that in closed systems kinetic energy was derived from the sum of potential and internal energies and vice versa. For a whole atmospheric column under hydrostatic equilibrium the ratio of $PE/IE = (c_p - c_v)/c_v$ is a constant, so the sum, $PE + IE = (PE/IE + 1)IE = (c_p/c_v)IE$, is

$$PE + IE = \frac{c_p}{g} \int_0^{p_{sfc}} T \, dp, \tag{8.16}$$

which is sometimes called the "total potential energy." Since it is related to $c_p T$, it is sometimes called the *sensible heat* (per unit mass) or sometimes *enthalpy* (see section 5.5).

Lorenz started out by considering available potential energy, AP, which he defined as the difference between the sum of the actual potential and internal energies and a theoretical minimum. The latter occurs when the atmosphere is horizontally stratified and statically stable. He estimated the available potential energy in the atmosphere, which is proportional to the spatial variance of temperature, to be in the order of 1/200th of the total potential energy. Also, he concluded that only about one-tenth of the available potential energy was converted into kinetic energy, which is related to the variance of wind speed.

Lorenz then used Reynolds' resolution to divide the energies into zonal and eddy components and to consider the energy flow through the system. The model is depicted in figure 8.1. Global area averages at constant pressure are represented by the overbar, e.g. $\overline{T}$, and deviations from the global mean by the prime, e.g. $T' = T - \overline{T}$. Zonal averages with respect to longitude, at constant latitude and constant pressure are donated by the square brackets, e.g. $[T]$, and deviations from them by the asterisk, $T^* = T - [T]$.

The rectangles represent the total energies. For example, the eddy available potential energy is given by

$$\overline{AE_e} = \frac{1}{2} \int_0^{\overline{p_0}} \frac{\overline{T^{*2}}}{(\Gamma_d - \overline{\Gamma})\overline{T}} \, dp, \tag{8.17}$$

where Γs are lapse rates. The diamonds represent energy flows and conversions. Again, as an example, the conversion from eddy available potential to eddy kinetic energy is given by:

$$C_e = -\frac{R}{g} \int_0^{\overline{p_0}} \frac{1}{p} \overline{T^* w^*} \, dp. \tag{8.18}$$

As indicated in section 8.2.3 there are different ways of taking averages so the equations may vary from one author to another. However, the general concept remains the same and the covariances in figure 8.1 give an idea of variables involved. To follow the energy through the system, the available potential energy is produced from uneven heating either in a S–N, GE_z, (1.2 × 10^5 J m^{-2}) or an E–W, GE_e, (0.7 × 10^5 J m^{-2}) orientation. Generation occurs as a positive covariance between temperature and heating, $[T]'[H]'$ and $T^* H^*$: warm air heated, cold air cooled. As is to be expected, most generation occurs between low and high latitudes but a significant amount is produced in the E–W contrasts. All of the zonally generated energy, and more, is converted by vertical and meridional winds to eddy potential energy to join that generated in

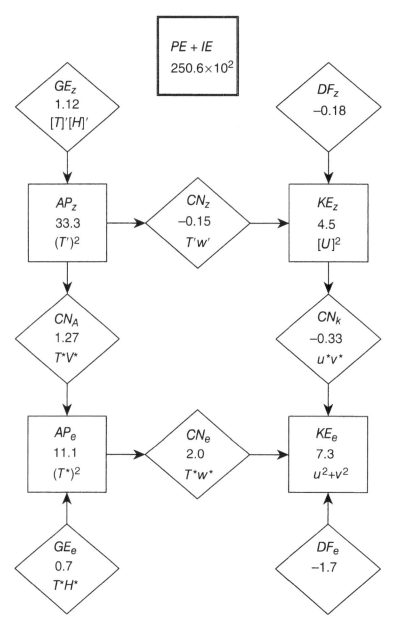

Figure 8.1 The energy cycle of the general circulation. The units of the energies (rectangles) are $\times 10^5\,\mathrm{J\,m^{-2}}$ and the units of the conversions (diamonds) are $\mathrm{W\,m^{-2}}$. The statistical terms are included to provide a general indication of the variables involved. Estimates are from Peixoto and Oort (1992)

the eddy domain. The conversion to kinetic energy (wind) takes place through the positive covariance of vertical motion in eddies, T^*w^* (rising warm air, sinking cold air) the most dominant of which are the extra-tropical cyclones and anticyclones. The eddy kinetic energy is dissipated through friction, DF_e, and converted to zonal kinetic energy, CN_k, by transfers up the scales of motion. Finally about half the zonal kinetic energy, KE_z, is dissipated, DF_z, and half converted to zonal potential energy (cold air rising, warm air sinking). This is surprising. As proposed by Hadley (1735) the overturning between low and higher latitudes was a *direct circulation*, potential to kinetic conversion. The kinds of estimates produced here suggest that, on the average for the globe, overturning in the meridional direction is *indirect*, i.e. kinetic to potential conversion.

The Lorenz model captures nicely the processes we observe in the atmosphere. For example, the extra-tropical cyclone–anticyclone pair participates a) in converting potential energy from the zonal to eddy domains through exchanging warm and cold air masses across latitude circles, and b) in converting potential energy to kinetic energy with warm air rising in the warm sector and cold air sinking especially behind the cold front. It also obviously transports and releases latent heat.

Within this scheme, momentum, sensible, and latent heat, and water vapor are transported and transformed. Usually their budgets are considered independently.

8.5 Heat Budget

We have already considered the earth–atmosphere as a whole and its radiative equilibrium temperature in section 6.4.2. Spatial uneven heating is, of course, what creates the available potential energy and eventually the circulation that redistributes that energy. Solar radiation receipt at the top of the atmosphere is a function of time of day, time of year, and latitude. To reduce the number of variables to two the daily sum is usually computed. This is what Milankovitch (1969) did early in the century to produce data like those used for figure 8.2. The earth–atmosphere system is full of feedback mechanisms. This set of theoretically obtained estimates is one of the few really independent inputs to that system. Therefore, if the solar output is assumed constant or can be predicted with reasonable accuracy this is a basic boundary condition for climate.

Houghton (1954) extended this analysis with newer radiation information to estimate the annual average absorption and emission. Today both the absorption and emission may be measured from satellites. Figure 8.3 is plotted from Houghton's theoretical data. It shows that latitudinally the earth–atmosphere is not in radiative equilibrium. Since less energy is emitted to space than is

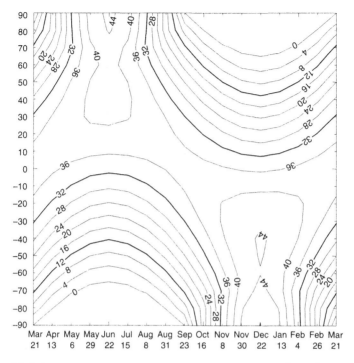

Figure 8.2 Solar receipt at the top of the atmosphere in MJ m^{-2} day^{-1} as a function of time of year and latitude. Data from List (1958)

absorbed from the sun in low latitudes and vice versa in high, it means that the circulation, which was established as a result of the equator–pole temperature gradient, transports heat. The amount estimated from the inequality by Houghton was 466×10^{18} J day^{-1} at 40°N. One result is the observed equator to pole temperature distribution. Table 8.4 displays the theoretical temperatures calculated by Milankovitch assuming a radiative balance. Also given are observed temperatures. Clearly, without atmospheric and oceanic circulations we should expect a larger temperature gradient than observed.

The paper already cited of Priestley (1949) provided the first calculations of the poleward flux of heat for the combined atmosphere and oceans. The latest empirical figure from Peixoto and Oort (1992) is 432×10^{18} J day^{-1} at 40°N, which is surprisingly close to Houghton's estimate.

More detailed estimates of the three-dimensional average diabatic heating of the atmosphere were performed by Newell et al. (1969). Some were reproduced in Peixoto and Oort (1992). They showed radiative cooling of the troposphere and warming of the tropical stratosphere; turbulent sensible heating of the lower atmosphere in the region of the sub-tropical highs; and latent heating

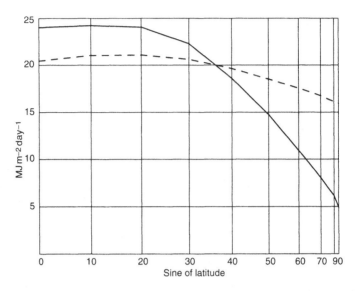

Figure 8.3 Annual mean solar radiation absorbed and long wave emission by the earth and atmosphere in $MJ\,m^{-2}\,day^{-1}$. Data from Houghton (1954)

Table 8.4 Theoretical and observed surface temperatures

Latitude	0	10	20	30	40	50	60	70	80	90
$T\,°C$, theoretical	32	31	28	22	14	4	−8	−20	−26	−28
$T\,°C$, observed	26	26	25	21	14	6	−1	−10	−17	

centered at 800–700 mb in mid-latitudes and 400–500 mb in the inter-tropical convergence zone.

With regard to a temporal and regional breakdown, Haines and Winston (1963) calculated and plotted the meridional transport of sensible heat across the whole 45°N latitude in the 850–500 mb layer at 5° longitude intervals for the period October 1958 to March 1962. The analysis displayed a strong annual cycle with preferred longitudes of flux. Those longitudes receiving warmer air might be expected to have above average temperatures compared to the latitudinal average. Also, the variabilities of the fluxes will appear in the variance of the temperatures.

Many geographical studies of the heat balance have been limited to the surface budget. For example, global surface heat fluxes were estimated by Budyko (1956). The processes and surface exchanges were the subject of a book by Sellers (1965) and they were the basis for the outline of a course by Miller (1968). Each contained good surveys of the literature.

8.6 Water Budget

Water must necessarily be included in the heat balance and it is a component of the studies already mentioned. However, it is important in its own right. Like the other fluxes accurate estimates are difficult to obtain. An early paper by Hutchings (1957) for water vapor divergence over England is a model for reviewing the accuracy of the elements involved in calculating the atmospheric flux. Much published research into energy flows concentrates upon the south–north transfers but, for water, west–east movement is just as important for divergence calculations. The source of water for precipitation is likely to come from any direction and that direction is likely to change with season. This is clearly demonstrated by Benton and Estoque (1954) who calculated the water transport over North America for different seasons in 1949, and by Howarth and Rayner (1991) for the southern hemisphere. Indeed, in the description and explanation of climate the fluxes of all variables from all directions are required.

Whereas precipitation is often thought to be well documented for land areas its magnitude and distribution over the oceans are poorly known. Evapotranspiration estimates are far more unreliable. Even the global average precipitation, which must equal the global average evapotranspiration, has not been accurately assessed. Peixoto and Oort (1992) list 973 mm yr^{-1} from Baumgartner and Reichel (1975) and 1,004 mm yr^{-1} from Sellers (1965). Some textbooks often give a significantly smaller magnitude. Estimates from continental areas are somewhat more accurate because additional information is available from runoff, the difference between precipitation and evapotranspiration.

8.7 Conversion Between Scales of Motion

The classical concept is that energy cascades down the scales of motion from the largest to the smallest. Thus it eventually appears as a temperature increase. L. F. Richardson recognized this in verse:

> Big whirls have little whirls
> That feed on their velocity
> And little whirls have lesser whirls
> And so on to viscosity – in the molecular sense. (Richardson, 1922, p. 66)

As Ashford (1985) points out, this is probably a parody of lines by Morgan, a nineteenth century mathematician, or possibly, as suggested by Hide (1983) from a poem by the Irish satirist and contemporary of Newton, Jonathan Swift. Ashford also proposed that, following Morgan, Richardson might have added:

> And the big whirls of bigger ones partake in the rotation,
> Until at last we reach the gen'ral circulation – in the global sense.

The fact that energy passes upwards from smaller to larger eddies in the atmosphere was demonstrated by Saltzman and Fleisher (1960a) and Saltzman and Teweles (1964). This work was based upon the spectral formulae developed by Saltzman (1957). He showed, following the work of Reynolds (1883) in the data domain, that in the spectral domain, kinetic energy transfer between waves of two different wave numbers was a function of the product of the amplitudes of three related wave numbers. For example, one of the terms in his Equation (47) for the time rate of change of the U component of kinetic energy, $\partial [|\, U(n)\,|^2]/\partial t$, is

$$\sum_{m=-\infty, m\neq 0}^{\infty} \left\{ \frac{in}{a\cos\phi} U(m)[v(-n)U(n-m) + \cdots] \cdots \right\},$$

where n and m are wave numbers. This says that the kinetic energy at wave number n is changed by the flux from or to wave number m but that it also involves wave number $n - m$. In other words, it is related to the bispectrum of the winds (Rayner, 1973a).

Barry Saltzman (1931–) was born in New York and completed a BS at NYU before going on for a masters and a doctorate (1954) at MIT. He has been a professor of geophysics at Yale since 1968.

Some early calculations on the conversions between the the zonal flow and the eddies were performed by Saltzman and Fleisher (1960b) and by Wiin-Nielsen et al. (1963), but more detailed analyses were done by Saltzman and associates at MIT. They found, for the nine year period from 1955 to 1964 for 15° to 80°N at 500 mb,

(1) that, in the mean, all waves ($n = 1-15$) transfer their energy to the zonally-averaged motion ($n = 0$) and, of more physical significance, the aggregate of all waves in the group $n = 2-15$ transfer energy to support the asymmetric polar vortex comprised of wave numbers zero and one, and (2) that, among the waves themselves, waves of $n = 2$ and 5–10 are sources of kinetic energy indicating a strong forced conversion of energy on the scale of the major continents and oceans. (Saltzman and Teweles, 1964)

The transfer of energy up the gradient is equivalent to having a *negative eddy viscosity*, the name given to this phenomenon by Starr (1968).

8.8 The General Circulation

The term *general circulation* refers to the complete description of atmospheric flow over the globe. It involves an investigation into the causes and the necessary energy supply. A comprehensive review was presented by Lorenz (1967) with a summary follow-up in Corby (1969). A thorough analysis of the transports and conversions was presented in a three volume work by Reiter (1969) and recent estimates have been included in the already referenced book by Peixoto and Oort (1992). The explanation of climate clearly involves the transfer of elements, where they occur, in what magnitudes, and in what frequencies. Consequently an understanding of the general circulation is a necessary precursor to this endeavor.

Chapter 9

Towards an Explanation of Climate

9.1 The Problem

The basic thesis of this book is that dynamic climatology seeks to explain climate. By explanation is meant the identification and ranking of the factors that control climate. It involves elucidating the physical mechanisms that relate a given factor to the variables of climate. To say that the precipitation amount in the US mid-west is related to the cycle of El Niño is not an explanation. El Niño itself must be explained and then the physical connection between the Pacific Ocean water temperature characteristics, including perhaps unrelated factors, and precipitation mechanisms must be thoroughly investigated. This is a monumental task and involves all aspects of research in the atmospheric and related disciplines. It goes beyond dynamic meteorology (Lorenz, 1996).

Whether the climate of interest is for a large region or for an individual location, most factors are expected to be global in nature. Also, even though there are millions of flapping butterfly wings, it is assumed that their net effect is random and small in differentiating climates. In contrast, factors like latitude, land and sea arrangements and relief distribution including subsurface ocean topography will dominate. However, since no real testing of this hypothesis has yet taken place, only future research will provide a solution.

Whatever the outcome, a series of prerequisite steps are necessary. They include the definition of climate, the selection of a method and the identification of inputs and constraints for the experiment.

9.1.1 Specification

The first step in the explanation must be the description of climate. It involves the characterization of weather systems over a wide range of scales in time and space, the subjects of most current introductory meteorology textbooks, and their synthesis with the necessary exhaustive set of statistics. It calls for the detailed four-dimensional measurement and analysis of these phenomena that was begun for extra-tropical cyclones and anticyclones by the Scandinavian meteorologists early in the twentieth century and continues for tornadoes and tropical cyclones at the end of the century.

While the description of individual and stereotype systems has advanced, the synthesis step has not yet been done and, indeed, it may not be possible. Two major questions exist.

How long a period of observation is necessary for the statistics to become stabilized?
Clearly climate is not a constant entity. The three major components of the system, the atmospheric composition, the surface characteristics of the earth, and energy input all display measurable variations.

1 The Medium, The Atmosphere
As we have seen in sections 5.3 and 6.4.4, even the composition of the medium with which we are dealing, the atmosphere, is continuously evolving in its natural state. However, that rate of change is usually assumed to be small over a few decades. Recent human activity, producing the well documented rapid increases in CO_2, CH_4, CFCs, etc., has obviously weakened this assumption. The question as to whether the system incorporating such increases can produce realistic stable climatic statistics remains unanswered. With regard to greenhouse gases, the role of water presents special problems. The amount, form, and location of water is very variable and, of course, is a function of the system. The residence time is short, perhaps the order of 10 days, and while the average is usually assumed to be constant, this does not take into account temperature trends. Volcanic effects, especially with regard to dust but also gases, appear to be transitory (2–3 years) for a particular eruption. The frequency of eruptions is a factor that needs to be considered.

2 The Earth
Earth size and rotation may be taken as constant. The rotation rate has decreased but not appreciably in the recent past. Also, although the position of the pole, the distribution of the continents and relief have changed significantly in geological time, they have varied little in these the last million years.

While the ocean is always recognized as having high inertia compared to the atmosphere its true effect is uncertain. Since currents are mainly produced by wind stress and salinity due to surface budgets they are a product of the system rather than being boundary conditions in themselves. Underwater topography changes do have to be taken into account in the long term. Surface land cover, which is a much shorter period phenomenon, has been the subject of much work in recent decades. Very little natural cover remains and it is unclear how much of that is in equilibrium with current climate. Built land cover is obviously an independently and rapidly varying factor.

3 Energy

The only independent boundary condition is solar receipt. As outlined in section 6.4.1 this varies through time but such variations can be, and are, incorporated into simulations. Solar absorption is controlled by the system as is the emission of long wave energy to space.

Landsberg and Jacobs (1951) gave a table of years necessary to produce stable statistics for different variables under different regional conditions. The maximum period suggested was 50 years. If we are in the midst of global warming then our observations are not stationary and no period will produce a stable set. Each of the controlling factors has a different period of existence that varies from seconds to millions of years although the assumption here is that the more important ones are long lived (10s of years). Climate, as synthesis of weather then, depends upon the interval selected for its definition. Depending upon that choice, the factors may be considered to be constant or varying. Varying factors clearly introduce fundamental problems because the statistics of climate may not become stabilized: the assumptions of stationarity, and therefore ergodicity, are violated. Perhaps future developments in statistics will allow successful analyses of such systems.

Is the atmospheric system *transitive*?

That is, for a set of initial conditions does only one set of statistics exist? Lorenz has investigated this problem in a number of papers starting in 1968 (Lorenz, 1968). He pointed out that some physical systems, such as dish-pan experiments, produce more than one set of statistics depending upon the conditions that happened to exist when the system became established. These are called *intransitive* systems. Such systems may switch from one set of statistics to another through, for example, an anomalous solar output. When the sun returns to its original conditions, the new climatic statistics may not coincide with the original ones.

Returning to ergodic theory, we note that certain transitive systems of equations may be converted into intransitive systems simply by changing the numerical value of a single constant. In the case of the dish-pan this constant might be the rate of rotation; for the atmosphere–ocean–earth system it might be a coefficient of turbulent viscosity or conductivity, whose most appropriate value in the atmosphere or ocean is uncertain in any case. If in a transitive system we do alter such a constant, but by an amount not quite enough to make the system intransitive, we may observe another form of behavior. Two particular

time-dependent solutions of the system may appear to have considerably different sets of statistics if the solutions are extended over only a moderate time span,
i.e., the system may appear to be intransitive. However, when the time span is
made sufficiently long, the solutions will be found to have similar statistics.
This means also that a single solution will exhibit different statistical properties
within different segments of a long time span. We have called systems of this
sort *almost intransitive*. (Lorenz, 1976)

Obviously, if the atmosphere is almost intransitive, additional methods of
dealing with the explanation of climate will have to be developed.

9.1.2 The Method

If we ignore these problems and can characterize climate by a set of statistics
then explanation involves relating those statistics to the influencing factors,
commonly called boundary conditions. Ideally they can be apportioned indi-
vidually and jointly and ranked according to their importance.

As indicated in chapter 1 there are three possible approaches to investigating how this system works: laboratory experiment and modeling; analytical
solution; and numerical modeling.

Laboratory modeling is very attractive and is very productive in fluid
dynamics applied to certain problems, such as to water moving through channels and to vehicles moving through water and air. It lends itself easily to
visualization and many features of the associated flows have been illustrated
by movies [see, for example, Shapiro (1961) and Shapiro (1972)]. Although
the atmosphere cannot be replicated in this way, dish-pan experiments by Fultz
(1951) and Hide (1953), have provided some valuable insights into rotating
flows. A review of the history and summary of the findings of this approach
is to be found in chapter VI of Lorenz (1967).

Analytical solution of the full system of equations listed in Chapter 7 appear
impossible for the foreseeable future. As Thompson (1961) pointed out

> the equations are simply too difficult to solve. In mathematical terms, the dif-
> ficulty is one of solving a general boundary- and initial-value problem for a
> system of six [Thompson did not include water] non-linear partial-differential
> equations in three dimensions. Even today, there are no known methods by which
> the solutions of such equations can be related explicitly to general boundary and
> initial conditions. (Thompson, 1961, p. 13)

The basic set of equations were available in the middle of the nineteenth
century and early analysis of them was reported in a series of papers by
von Helmholtz (1858). Drastic simplification of the equations do yield results
in limited situations. For example, Queney (1948), Scorer (1949), Scorer and

Wilkinson (1956) and Crapper (1959) were able to specify the air flow over a hill of simple form. More recently, Zehnder (1991) was able to show that vorticity centers develop downstream of the Sierra Madre Mountains in Mexico which might account for the high incidence of tropical cyclones in the Eastern N. Pacific.

Numerical solution, however, is the only viable alternative for the larger system.

9.1.3 Requirements

Several prerequisites are necessary for a numerical solution:

Defining Relationships

The specific form of the equation set must be selected and converted to numerical form. Typically the spherical polar coordinates with some kind of modified vertical coordinate are used (see section 7.11). These are then written in a finite difference form (section 2.8). Often for global simulation, some procedures produce better results if the equations are converted into the spectral domain so these transformation algorithms may be required.

Although these are the basic set of equations there are others which must be added to make them solvable numerically. For example, the radiation and turbulent flows need to be defined.

Grid and Scale

Part of the last step involves the selection of the grid form and interval both in the vertical and horizontal. Obviously a simple rectangular or latitudinal grid will not be applicable in polar regions for a global model. In the vertical a higher resolution is typically applied at lower levels. Where the grid interval varies, scale factors must be included.

Boundary Conditions

A major problem in climatic simulation, as in weather forecasting, is the incorporation of boundary conditions. These involve all the factors mentioned in section 9.1.1. The receipt of solar radiation is a relatively simple issue. The absorption is not, since it depends upon factors like cloud amount, water vapor content and surface albedo, all of which are a function of the system. For example, the continental surface wetness, which is due to surface condensation, precipitation, and evapotranspiration affects the albedo, conductivity, specific heat, evapotranspiration, etc. Over time it also affects plant growth, which in turn affects all those elements and, in addition surface roughness. In other words, the feedback processes and non-linearity are highly complex.

Parameterization

Processes in the atmosphere occur at all scales. In section 8.7 we saw how energy moves both up and down this continuum. Certain scales contain more energy than others. Vinnichenko (1970), for example, has shown that over time the kinetic energy is concentrated in daily-weekly periods accompanying a strong peak at the annual cycle and smaller peaks at one day and one minute. At the spatial scales global and synoptic systems dominate but significant and critical processes act at the micro level. Even though the physics may be well understood at the smaller temporal and spatial intervals it is usually impractical to simulate them on a regional or global scale. Consequently their resultant effects are generalized. This is called *parameterization*. At the end of the twentieth century this subject is a very active area of development.

Flux Constraints

The simulation must not only produce idealistic atmospheric systems it must also represent the various fluxes accurately. A simple, yet not always fulfilled requirement, is that the mass of the atmosphere remains constant. Similar checks and balances must be observed for other systems such as the oceans where, in early models, salt was not conserved.

Existing Climate and Validation

The first simulation for a climatic model must be based upon existing conditions, usually known as a *control run*. The output must then be compared to observed climate to see how well the model works. As we have seen, because climate has not been well defined, because it may not be transitive, and because it may not be stationary, the comparison is problematic. Numerical models today show large discrepancies even with mean observed magnitudes, especially on a regional basis. Obviously if the output of the model only displays partial correlation with observed data, the value of any manipulation with boundary conditions, etc., to test their significance, will be seriously weakened. That is the current state of climate models. It has often led to very different estimates of warming that are expected to accompany the increase of greenhouse gases. It is surprising that they even display the same overall sign (warming) although from region to region they present more divergent estimates.

Regional variation makes validation difficult.

9.2 Numerical Modeling

9.2.1 Early Development

According to Cressman (1996) the first attempt at atmospheric numerical computation was performed by Exner and Defant about 1908. V. Bjerknes had already conceived of "A Rational Method for Weather Prediction" using graphical methods that he outlined in a lecture in 1903 (Friedman, 1989). However, Richardson is usually credited with the first full calculations. His book, *Weather Prediction by Numerical Process*, explaining the methods he used between 1910 and 1918 is a classic, not only for the methods and calculations, but also for its vision for the future.

Lewis Fry Richardson (1881–53) was the youngest of seven children in a Quaker family who lived in Newcastle in NE England. He became interested in science as a child and kept a detailed diary of natural history including the weather. At Cambridge he studied both physics (under J. J. Thomson) and biology. One of his first jobs was in a peat company where he was concerned with the flow of water in the peat-moss. Initially he used graphical methods but later applied finite differences to solve his differential equations (see section 2.8). Richardson had several jobs before being appointed Superintendent at Eskdalemuir Observatory by the Meteorological Office. It was there that, among other researches, he completed much of the theoretical part of his book (Ashford, 1985). Being a conscientious objector he was granted exemption from military service when conscription for the First World War was introduced in 1916. Despite this he had already applied for leave, which was denied, to join the Red Cross in 1914. In 1916 he resigned his observatory post and joined the Friends (Quaker) Ambulance Unit. During Richardson's service in France he completed a practical example for his book. At the end of the War, because of his resignation he could not be automatically re-appointed to Eskdalemuir, and he applied to Shaw, who was Director of the Meteorological Office, for support. He was awarded a year's salary to work at the Benson Observatory that was run by W. H. Dines. He resigned that post and visited Bjerknes in Bergen before taking a lecturing job at Westminster Training College. During this period Richardson completed his book and published papers on turbulence (see section 7.6.4) and psychology. Not only did he obtain a DSc from London University in 1925 based upon his published papers, the same year he also passed examinations in psychology and in pure and applied mathematics at Cambridge (Ashford, 1985). In 1929 he was appointed Principal of Paisley Technical College. He retired in 1940, moving to Kilmun near Glasgow in 1943 where he continued his researching and publishing in meteorology and psychology until his death.

For his numerical forecast Richardson used equations very similar to those presented in previous chapters. The equations of motion in spherical polar coordinates he took from Lamb's 4th edition. Much of the text is taken up with the consideration of radiation, cloud processes, turbulence, and surface and subsurface processes. These are the very topics which are at the center of a great deal of research still, 80 years later. Unfortunately, due to a number of factors, Richardson's forecast produced an unacceptable result – a pressure increase of "about 60 millibars in six hours" (Richardson, 1922, p. 212). Nevertheless this was a fantastic accomplishment especially as his techniques are essentially the same as those still in use.

In some ways even more remarkable was his vision for the future of weather forecasting. Whereas his "computers" were people, he envisioned the parallel machine:

After so much hard reasoning, may one play with a fantasy? Imagine a large hall like a theatre, except that the circles and galleries go right round through the space usually occupied by the stage. The walls of this chamber are painted to form a map of the globe. The ceiling represents the north polar regions, England is in the gallery, the tropics in the upper circle, Australia on the dress circle and the antarctic in the pit. A myriad computers are at work upon the weather of the part of the map where each sits, but each computer attends only to one equation or part of an equation. The work of each region is coordinated by an official of higher rank. Numerous little "night signs" display the instantaneous values so that neigbbouring computers can read them. Each number is thus displayed in three adjacent zones so as to maintain Communication to the North and South on the map. From the floor of the pit a tall pillar rises to half the height of the hall. It carries a large pulpit on its top. In this sits the man in charge of the whole theatre; he is surrounded by several assistants and messengers. One of his duties is to maintain a uniform speed of progress in all parts of the globe. In this respect he is like the conductor of an orchestra in which the instruments are slide-rules and calculating machines. But instead of waving a baton be turns a beam of rosy light upon any region that is running ahead of the rest, and a beam of blue light upon those who are behindhand. (Richardson, 1922, p. 219)

9.2.2 The Electronic Computer

The development of the electronic computer during the 1940s inspired a number of individuals to investigate their application. In 1946 the mathematician John von Neumann (1903–57) of The Institute for Advanced Study at Princeton, NJ, seized upon "an investigation of the theory of dynamic meteorology" as the objective of his proposal to the US Navy Department to establish a Meteorology Project (Thompson, 1983).

Johann von Neumann was born in Budapest, Hungary, the oldest of three sons of a well-to-do Jewish family. He was privately educated and by age 19 was already recognized as a professional mathematician. He was invited to a visiting lectureship at Princeton in 1930 and obtained a permanent professorship there in 1931 and moved on to the Institute for Advanced Studies two years later. He was consultant for the atomic bomb development at the Los Alamos but his significant contribution from our standpoint is his influence on the evolution of computers especially on the development codes by which a fixed system of wiring could solve a great variety of problems (Ulam, 1958).

In the years following 1946 a large number of eminent meteorologists both from the US and overseas, listed by Thompson, were attracted to the project. Of particular note were Charney, Eliassen, and Fjörtoft. Charney (1948, 1949) provided the the necessary simplified equations which were adapted for, what appears today, as a very primitive computer. Because the Princeton machine was not operational in April 1950 the first test supervised by Platzman, Smagorinsky, and Freeman, was run on the Electronic Numerical Integrator and Calculator (ENIAC) at the Aberdeen Proving Ground 30 miles NE of Baltimore. Although the model was relatively simplistic and the prediction not particularly good, the experiment was considered to be an enormous success (Charney et al., 1950).

Jule Charney (1917–81) stands out as one of the most significant figures in meteorology. Most people will point to the papers referenced above but, as illustrated by Phillips (1982), his influence goes well beyond those contributions. Jule was born on New Year's Day in San Francisco of Russian Jewish heritage. The name Charney is apparently the Polish pronunciation of "black" referring to the dark skin of the family (Platzman, 1987). As a child he lived in Los Angeles before he moved at 14 to New York with his mother when his parents separated. It was there that he became interested in mathematics and later science. By 1934 his parents reunited and he had returned to California where he attended UCLA, graduating with a BA in 1938 and a mathematics MA in 1940. As a graduate student he heard about meteorology from Holmboe, and was exposed to the work of the Bjerknesses and went to a lecture by von Kármán at Cal Tech. Uncertain as to his future career the young Charney asked von Kármán for advice. He recommended the new field of meteorology. As a result Charney interacted with meteorologists such as J. Bjerknes, Lieutenant Philip Thompson, and with Neiburger and Fletcher both of whom had been students of Rossby and from whom he heard about Rossby's 1939 paper. He was awarded his doctorate in 1946 and obtained funding to work with Solberg in Oslo. On his way there he visited Rossby in Chicago and Thompson who was by then at Princeton. Charney held an appointment at Princeton from 1948 to 1966 when he was appointed Sloan Professor of Meteorology at MIT.

Following the ENIAC experiment, development occurred quickly as Charney moved from a one- through a two- to a three-level quasi-geostrophic model. The new von Neumann computer came online in 1952 and work proceeded simultaneously in Sweden, Norway, and England. Under the leadership of Rossby, Sweden was the first country to make numerical prediction operational in December 1954. The US followed in May 1955. Baroclinic models were introduced in the early 1950s but models applying the primitive equations, the ones used by Richardson, were not practical until computing power increased in the 1960s. The skill of the forecasts finally surpassed traditional methods in about 1965 when routine numerical prediction using primitive equation models was implemented in the US (Cressman, 1996).

At the end of the twentieth century most developed countries run numerical forecast models. Many universities and small companies also produce daily forecasts using global or, more likely, meso-scale models that abstract their initial and, sometimes, boundary conditions from national models.

9.3 Climate Modeling

Although the initial numerical models were developed for short term prediction, climatologists immediately saw a potential in them for investigating climates in a way never before possible. For example, by changing the boundary conditions the weather and circulation during past climates could be researched. One early attempt at this approach involving both meteorologists and geographers was performed by Williams et al. (1974) where continental glaciers were introduced.

An enormous number of projects like this, known as *sensitivity studies* have been performed and their results published. The research continues at a rapid rate. They range over a wide variety of modified boundary conditions such as solar radiation, albedo, surface cover, thermal pollution, volcanic dust, and relief. Global warming estimates based on the increase of greenhouse gases are of the same genre. An example of the way in which these major factors may be studied is provided by Manabe and Broccoli (1990). These authors found that the low soil moisture of continental interiors was controlled more by the rainshadow effect than by the usual explanation, distance from the ocean. The conclusions of such studies depend very much upon the validity of the model components that are used. Improvements are being made continuously and duplication of experiments upgrade results and reduce their statistical confidence bands.

Individuals wishing to participate need to gain a thorough knowledge of the physics as well a broad understanding of the societal repercussions of publicizing any conclusions drawn.

Throughout this book it has been emphasized that only introductory material has been presented. A large literature exists on each small component. For example, early books by Richtmyer and Morton (1967) and Haltiner and Williams (1980) and several GARP publications (WMO, 1972, 1973, 1976) may be consulted for basic numerical procedures. Good early surveys of climate simulation were given by Washington and Parkinson (1986) and in a volume commemorating the retirement of Joseph Smagorinsky as Director of the Geophysical Dynamics Laboratory (GFDL) of NOAA (Manabe, 1985). A more recent comprehensive review is given in the 800 page volume edited by Trenberth (1992). For the novice interested in pursuing this line of research the book by McGuffie and Henderson-Sellers (1997), which contains a computer disk of programs, is a good place to start. A source for global warming and simulation results at the end of the twentieth century is the US Global Change Research Office (GCRIO). Unfortunately material in books is often obsolete before it is printed. Having learned the background the only way a researcher can become current is to attend seminars by those practicing the craft and to interact on a daily basis with as many model development groups as possible. Numerous web sites relay the latest developments. The field of dynamic climatology is truly a dynamic one: it is not stationary.

Chapter 10

Concluding Remarks

The aim of this volume is to present in a chronological and coherent way the basic relationships that have been developed to describe the atmosphere. These are already available as introductory matter in many texts so it breaks no new ground. Indeed, no current research material has been included and important topics such as parameterization, cloud physics, and atmospheric chemistry, have been omitted. These are the subject of current courses and seminars in the atmospheric sciences. Interested persons are urged to pursue each topic in the literature and in formal courses.

One way to gain a thorough understanding of specific relationships and algorithms is for the student to follow the computer code line by line. It is a slow process but it is often far more informative than the original equations and the narrative descriptions alone. Only a few organizations make their source codes available to the general public. One is the National Center for Atmospheric Research (NCAR), a facility managed by the University Corporation for Atmospheric Research (UCAR) in Boulder, Colorado. Funded partly by the US National Science Foundation this research unit makes its global and mesoscale models available over the internet.

For those historically inclined there is a wealth of biographical material available. For students interested in the more technical material there is no better source than the original research papers and books. The task of finding such documents is being made easier because significant papers and the works of many of the notable scientists are being published as collections (Saltzman, 1962). Even the lazy among us, who do not want to struggle with a foreign language, can find excellent translations of many fundamental papers (Boissonnade and Vagliente, 1997).

Theory and observation must go hand in hand. Climate as the synthesis of weather builds upon an understanding of the instantaneous state of the atmosphere. Each locality is influenced to varying degrees by different weather phenomena. The frequency and peculiarities of these combine to create climate. Only through the documentation and explanation of the whole range of weather systems can we hope to develop a theory of climate, a true dynamic climatology. Simulation models are the the most viable approach to providing answers at this time but they are not the solution in themselves. They only represent the current (usually past) state of theory and observation. The models must be made to represent the reality that we construct as faithfully as possible and their output must always be carefully compared with any relevant observational data that exist.

As stated in the Preface, this introductory survey is meant to provides a basis for those who are interested in finding out what some of the basic underlying principles of climate are and how they developed.

Tor Bergeron called for a truly dynamic climatology. That is what it has become. The lines between meteorology and climatology have blurred. Once climatology was seen as nothing more than "a compiling of statistics." Now it encompasses the whole of meteorology and goes beyond: to explain climate not only must we first explain the instantaneous state we must also explain the long term sequence of states. Because of its very broad nature it must include the fluid dynamics theoretician, the numerical analyst, the statistician, the atmospheric observer, and more. It remains to be seen whether the twenty-first century will produce explanation of climates in terms of the factors that control them. Then true climatic prediction may be possible and the social and behavioral consequences may be dealt with in a rational manner.

Appendix A

Power Notation

Prefix name	Symbol	Power	Magnitude
nano	n	10^{-9}	0.000000001
micro	μ	10^{-6}	0.000001
milli	m	10^{-3}	0.001
		10^{0}	1.0
kilo	k	10^{3}	1000.
Mega	M	10^{6}	1000000.
Giga	G	10^{9}	1000000000.
Tera	T	10^{12}	1000000000000.
kibi	*k*	2^{10}	1024.
Mebi	*M*	2^{20}	1048576.
Gibi	*G*	2^{30}	1073741824.

For binary notation see *Science* vol. 283, March 12, 1999, p. 1631.

Appendix B

Constants

Name	Magnitude	Units
Avogadro number [N_0]	6.022529×10^{26}	$kmole^{-1}$
Boltzmann constant [k]	1.380546×10^{-23}	$J K^{-1}$
Exponential number [e]	2.718381	
Gas constant [R^*]	8.31434×10^3	$J kmole^{-1} K^{-1}$
Gas constant – dry air [R_d]	287	$J kg^{-1} K^{-1}$
Gravitational constant [G]	6.6705×10^{-11}	$N m^2 kg^{-2}$
Light year	9.460529×10^{12}	km
	63.2×10^3	sun–earth dist
Pi [π]	3.14159265	
Planck constant [h]	$6.6255916 \times 10^{-34}$	$J s$
Speed of light [c]	2.997925×10^8	$m s^{-1}$
Stefan–Boltzmann constant [σ]	5.6697×10^{-8}	$W m^{-2} K^{-4}$
Volume ideal gas (*std. T, p*)	22.4136	$m^3 kmole^{-1}$
Specific heats		
Water	4190	$J kg^{-1} K^{-1}$
Ice	2137	$J kg^{-1} K^{-1}$
Dry Air [c_p]	1005	$J kg^{-1} K^{-1}$
Wet sandy soil	1257	$J kg^{-1} K^{-1}$
Dry sandy soil	838	$J kg^{-1} K^{-1}$
Conductivity [λ]		
Water	0.629	$J K^{-1} m^{-1} s^{-1}$
Air	0.0210	$J K^{-1} m^{-1} s^{-1}$
Rock	4.6	$J K^{-1} m^{-1} s^{-1}$

(*Continued*)

Name	Magnitude	Units
Density [ρ]		
Water	10^3	$kg\,m^{-3}$
1 kg on 1 m^2 is 1 mm deep		
Air	1.22	$kg\,m^{-3}$
Fir	0.5×10^3	$kg\,m^{-3}$
Rock	2.7×10^3	$kg\,m^{-3}$
Latent heat of water at 0°C		
vaporization [l_v]	2.5008×10^6	$J\,kg^{-1}$
sublimation [l_s]	2.8345×10^6	$J\,kg^{-1}$
fusion [l_w]	0.3337×10^6	$J\,kg^{-1}$
Sun		
Solar radius	698×10^3	km
Mass	2.0×10^{30}	kg
Distance to nearest star		
(Proxima in Alpha Centauri)	4.3	light years
	271×10^3	sun–earth dist
Solar constant	1376	$W\,m^{-2}$
Earth		
Radius (Average)	6371	km
(polar)	6356.9	km
(equator)	6378.1	km
Area (Total)	510×10^6	km^2
Volume	$1083319.78 \times 10^{24}$	km^3
Earth mass	5.975×10^{24}	kg
Atmospheric mass	5.15×10^{18}	kg
Angular velocity	7.292116×10^{-5}	s^{-1}
Equator linear velocity	465.1	$m\,s^{-1}$
Acceleration of gravity, g	9.80665	$m\,s^{-2}$
Obliquity of ecliptic (Tilt)	23.45	°
Mean orbital radius	149.68×10^6	km
Mean solar day	86400	s
Mean sidereal day	86164.1	s
Mean solar year	365.2422	days
	31.5569×10^6	s
Mean orbital velocity	29.77	$km\,s^{-1}$

Appendix C

Conversions

Many of the following were taken from List (1958)

Conversion	From	To	Multiply by
Length	inches	m	0.0254
	feet	m	0.3048
	miles	m	1609.344
	naut. mile	m	1853
Area	acre	m^2	4047
	hectare	m^2	10000
	in^2	m^2	0.00064516
	ft^2	m^2	0.09290304
	$mile^2$	m^2	2.58999×10^6
Volume	in^3	m^3	16.3871×10^{-6}
	ft^3	m^3	0.0283168
	liter	m^3	0.001
	fluid oz (US)	m^3	29.5735×10^{-6}
	fluid oz (UK)	m^3	28.413×10^{-6}
	gallon (US)	m^3	3.78541×10^{-3}
	gallon (UK)	m^3	4.546×10^{-3}
Speed	knot	$m\,s^{-1}$	0.514791
	mph	$m\,s^{-1}$	0.44704
	$ft\,s^{-1}$	$m\,s^{-1}$	0.3048
	$ft\,min^{-1}$	$m\,s^{-1}$	5.08×10^{-3}
	$km\,hr^{-1}$	$m\,s^{-1}$	0.277778

(Continued)

Conversion	From	To	Multiply by
Mass	ounce	kg	0.0283495
	pound	kg	0.4535923
	2000 lb	kg	907.1846
	2240	kg	1016.047
Pressure	in (Hg)	Pa	3386.39
	mb	Pa	100
	lb in^{-2}	Pa	6894.76
Force	lb-wt	N	4.44822
Energy	kW-hr	J	3.6×10^6
	Btu ($^\circ$ F)$^{-1}$ lb^{-1}	J	1055.07
	erg	J	10^{-7}
	cal	J	4.19
	kcal = Food cal	J	4190
Power	horsepower	W	746
	Btu min^{-1}	W	16.75
	cal min^{-1}	W	0.0698
Energy/area	langley	J m^{-2}	41.9×10^3
Power/area	langley min^{-1}	W m^{-2}	698
	kly yr^{-1}	J m^{-2} yr^{-1}	41.9×10^6

Appendix D

World Data

Since we usually live on land our view of the world tends to be biased towards the continental areas, only 29 percent of the earth's surface. Also, since most developed countries are in mid-latitudes their people tend to think of the earth in terms of mid-latitude landscapes. Only 13 percent of the earth's surface is to be found between 30 and 50°N and some of that is ocean. Half the earth's surface is to be found between 30°N and 30°S. The atmosphere is affected by the whole surface.

Also listed below are some of the larger countries in rank order of size. Mid 1999 population estimates are from the Population Reference Bureau, Inc.

Name	Area (km^2)	Population ($\times 10^6$)
Earth	510×10^6	
Oceans (71%)	361×10^6	
Continents (29%)	149×10^6	5982
Oceans		
Pacific	166×10^6	
Atlantic	83×10^6	
Indian	74×10^6	
Arctic	14×10^6	
Continents		
Asia	44.8×10^6	3637
Africa	30.3×10^6	771

(*Continued*)

Name	Area (km^2)	Population ($\times 10^6$)
N. America	24.3×10^6	475
S. America	17.8×10^6	339
Antarctica	13.2×10^6	
Europe	9.9×10^6	728
Australia & Oceania	8.5×10^6	30
Lakes		
Caspian	371×10^3	
Superior	82×10^3	
Victoria	69×10^3	
Huron	59×10^3	
Michigan	58×10^3	
Aral	41×10^3	
Countries		
Old USSR	17.3×10^6	240
Canada	10.0×10^6	31
China	9.6×10^6	1254
US	9.5×10^6	273
Brazil	8.5×10^6	168
Australia	7.7×10^6	19
India	3.2×10^6	987

Bibliography

Abbe, C. 1906: Benjamin Franklin as meteorologist. *Proceedings of the American Philosophical Society*, **45**, 117–28.

Abramowitz, M. and Stegun, I. A. (eds) 1964: *Handbook of Mathematical Functions with Formulas, Graphs and Mathematical Tables*. Washington, DC: US Government Printing Office.

Agassiz, L. 1967: *Studies on Glaciers*. New York: Hafner Publishing McGraw-Hill Book Co. [These are translations by A. V. Carozzi of Agassiz's "Discours de Neuchâtel" of 1837 and "Études sur Les Glaciers" of 1840.]

AMS, 1974: SI Units to be used in AMS journals. *Bulletin of the American Meteorological Society*, **55**, 926–30. Statement by the American Meteorological Society.

Aris, R. 1962: *Vectors, Tensors, and the Basic Equations of Fluid Mechanics*. Englewood Cliffs, NJ: Prentice-Hall, Inc.

Arrhenius, S. 1896: On the influence of carbonic acid in the air upon the temperature on the ground. *The Philosophical Magazine*, **Ser 5, vol. 41**, 237–76. Also published in Swedish in *Bihang till Kungl. Svenska Vetenskapsakademiens handlingar*, **22**, 1–102. and see Feb 77 issue of *Ambio*.

Ashford, O. M. 1985: *Prophet – or Professor: The Life and Work of Lewis Fry Richardson*. Bristol and Boston: Adam Hilger Ltd.

Atkinson, B. W. 1981: *Meso-scale Atmospheric Circulations*. Boston, MA: Academic Press.

Barber, N. F. 1961: *Experimental Correlograms and Fourier Transforms*. New York: Pergamon Press.

Baumgartner, A. and Reichel, E. 1975: *The World Water Balance*. Amsterdam: Elsevier.

Beckmann, P. 1970: *A History of π (pi)*. Boulder, CO: The Golem Press.

Benton, G. S. and Estoque, M. A. 1954: Water vapor transfer over the North American continent. *Journal of Meteorology*, **11**, 462–77.

Berger, A. 1977: Long-term variations of the earth's orbital elements. *Celestial Mechanics*, **15**, 53–74.

Bergeron, T. 1930: Richtlinien einer dynamischen Klimatologie. *Meteorologische Zeitschrift*, **47**, 246–62.

Bergeron, T. 1935: On the physics of cloud and precipitation. In *Procès verbaux des séances de l'Association de Metiorolgie*. UGGI, 5e assemblée générale, Lisbonne, Sept. 1933.

Bergeron, T. 1959: The young Carl-Gustaf Rossby. In Bolin (1959) 51–5.

Berry, F. A., Bollay, E., and Beers, N. R. (eds) 1945: *Handbook of Meteorology*. New York: McGraw-Hill Book Co.

Bjerknes, J. 1919: On the structure of moving cyclones. *Monthly Weather Review*, **47**, 95–9.

Bjerknes, V. 1898: Über die Bildung von Cirkulationbewegungen und Wirbeln in reibungslosen Flüssigkeiten. *Skrifter udg. af Videnskabsselskabet i Christiania*, **5**, 29.

Black, T. L. 1994: The new NMC mesoscale Eta model: Description and forecast examples. *Weather Forecasting*, **9**, 265–78.

Blackman, R. B. and Tukey, J. W. 1958: *The Measurement of Power Spectra from the Point of View of Communications Engineers*. New York: Dover Publications, Inc.

Bohren, C. B. and Albrecht, B. A. 1998: *Atmospheric Thermodynamics*. Oxford and New York: Oxford University Press.

Boissonnade, A. and Vagliente, V. N. (eds) 1997: *Analytical Mechanics: Translation of Lagrange's Mécanique Analytique*. Boston, MA: Kluwer Academic Publishers. Boston Studies in the Philosophy of Science, vol. 191.

Bolin, B. (ed.) 1959: *The Atmosphere and Sea in Motion: Scientific Contributions to the Rossby Memorial Volume*. New York: The Rockefeller Institute Press in association with Oxford University Press.

Bolin, B. 1999: Carl-Gustaf Rossby. The Stockholm period 1947–1957. *Tellus*, **51 A-B**, 4–12.

Bouger, P. 1729: *Essai d'optique sur la graduation de la lumière*. Paris. [Expanded version, "Essai" replaced by "Traité" published posthumously in 1760, translated by Middleton (1961).]

Boyer, C. B. 1939: *The Concepts of Calculus*. New York: Hafner Publishing McGraw-Hill Book Co.

Boyle, R. 1662a: New Experiments Physico-mechanical, Touching the Spring of the Air and its Effects. [2nd edition appendix, not in 1st edition in 1660.]

Boyle, R. 1662b: Of the mechanical origin, or production of heat. Reprinted from Works of the Honorable Robert Boyle, vol 4, London, 1772, 244–51. [No original date given.] In Lindsay (1975).

Brillinger, D. R. 1975: *Time Series: Data Analysis and Theory*. New York: Holt, Rinehart and Winston, Inc.,

Brillinger, D. R. and Rosenblatt, M. 1967: *Computation and Interpretation of k-th Order Spectra*. In Harris (1967).

Brooks, F. A. 1960: *An Introduction to Physical Meteorology*. Davis, CA: University of California.

Brothers, H. J. and Knox, J. A. 1998: New closed-form approximations to the logarithmic constant e. *Mathematical Intelligencer*, **20** (5), 25–9.

Brown, R. A. 1991: *Fluid Mechanics of the Atmosphere*. New York: Academic Press, Inc.

Brown, S. C. 1981: Benjamin Thompson (Count Rumford). In Gillispie (1981).

Brown, T. J., Berliner, L. M., Wilks, D. S., Richman, M. B., and Wilke, C. K. 1999: Statistics education in the atmospheric sciences. *Bulletin of the American Meteorological Society*, **80**, 2087–97.

Brush, S. G. 1981: Ludwig Boltzmann. In Gillispie (1981).

Budyko, M. I. 1956: *Teplovoi balans zemnoi poverkhnosti*. Gidrometeorologicheskoe Izdatel'stvo. [English translation by N. A. Stepanova 1958: The heat balance of the earth's surface. Office of Technical Services, US Dept. of Commerce, Washington, DC.]

Byers, H. R. 1959: Carl-Gustaf Rossby, the organizer. In Bolin (1959) 56–9.

Cajori, F. 1893: *A History of Mathematics*. New York: Macmillan Company.

Cajori, F. 1929: *A History of Mathematical Notations*. Chicago, IL: The Open Court Publishing Company.

Carnot, L. 1803: *Principes fondamentaux de l'équilibre et du mouvement*. Paris: Imprimerie De Crapelet. [Extract translated in Lindsay (1975).]

Carnot, S. 1824: *Réflexions sur la puissance motrice du feu et les machines propres à d'evelopper cette puissance*. [Reflections on the motive power of fire and on machines fitted to develop that power.] In Mendoza (1977).

Carslaw, H. S. and Jaeger, J. C. 1959: *Conduction of Heat in Solids*. Oxford: Clarendon Press.

Charney, J. G. 1945: Radiation. In Berry et al. (1945) chapter IV, 283–311.

Charney, J. G. 1948: On the scale of atmospheric motions. *Geophysiske Publikasjoner*, **17**, 3–17.

Charney, J. G. 1949: On the basis for numerical prediction of large scale motion in the atmosphere. *Journal of Meteorology*, **6**, 371–85.

Charney, J. G., Fjörtoft, R., and von Neumann, J. 1950: Numerical integration of the barotropic vorticity equation. *Tellus*, **2**, 237–54.

Clapeyron, E. 1834: *Memoir on the Motive Power of Heat*. In Mendoza (1977).

Clausius, R. 1850: *On the Motive Power of Heat, and on the Laws which can be Deduced from it for the Theory of Heat*. [Ueber die bewegende Kraft der Wäme und die Gesetzewelche sich daraus für die Wärmelehre selbst ableiten lassen, published in *Annalen der Physik*, **79**, 368–97, 500–24.] In Mendoza (1977).

Clausius, R. 1875: *The Mechanical Theory of Heat*. London: Macmillan and Co. [Translated by Walter R. Browne and published in 1879.]

Collins, G. W. 1989: *The Foundations of Celestial Mechanics*. Tuscon, AZ: Pachart Publishing House. [Astronomy and Astrophysics Series, 16.]

Corby, G. A. (ed.) 1969: *The Global Circulation of the Atmosphere*. Royal Meteorological Society, London, England. Papers of a joint conference of the Royal Meteorological Society, the American Meteorological Society, and Canadian Meteorological Society, August 1969.

Cornish, E. A. and Fisher, R. A. 1937: Moments and cumulants in the specification of distributions. *Extrait de la Revue de l'Institut International de Statistique*, **4**, 1–14. [Also published in Fisher (1950) as paper #30.]

Crapper, G. D. 1959: A three-Dimensional solution for waves in the lee of mountains. *Journal of Fluid Mechanics*, **6**, 51–76.

Cressman, G. P. 1996: The origin and rise of numerical weather prediction. In Fleming (1996), 21–39.

Crichton, M. 1990: *Jurassic Park*. New York: Ballantine Books.

Croll, J. 1864: On the physical cause of the change of climate during geological epochs. *Philosophical Magazine*, **Ser 4 vol. 28**, 121–37.

Crosland, M. P. 1981a: Amedeo Avogadro. In Gillispie (1981).

Crosland, M. P. 1981b: Joseph Louis Gay-Lussac. Gillispie (1981).

Currie, R. G. 1988: Climatically induced cyclic variations in United States crop production: implications in economic and social science. In Erickson and Smith (1988), 181–241.

d'Alembert, J. L. R. 1743: The vis viva controversy. In Lindsay (1975), 135–8.

Daub, E. E. 1981: Rudolf Clausius. In Gillispie (1981).

Davidson, B., Friend, J. P., and Seitz, H. 1966: Numerical models of diffusion and rainout of straospheric radioactive materials. *Tellus*, **18**, 301–15.

de Saint-Venant, A. J. C. B. 1843: Unknown title. *Comptes Rendus de l'Academie des sciences*, **xvii**, 1240.

Defrise, P. 1964: Tensor calculus in atmospheric mechanics. In Landsberg and van Miegham (1964), 261–15.

Descartes, R. 1644: *Principia philosphiae*. Extract translated in Lindsay (1975).

Drabin, I. and Drake, S. 1960: *Galileo on Motion and on Mechanics*. Madison, WI: University of Wisconsin.

Draper, N. R. and Smith, H. 1966: *Applied Regression Analysis*. New York: John Wiley & Sons, Inc.

Durst, C. S. 1951: Climate – the synthesis of weather In Malone (1951), 967–75.

Dutton, J. A. 1976: *The Ceaseless Wind: An Introduction to the Theory of Atmospheric Motion*. New York: McGraw-Hill Book Co.

Einstein, A. 1905: Über einen die Erzeugung und Umwandlung des Lichtes betreffenden heuristischen Standpunkt [On a heuristic viewpoint concerning the generation and transformation of light]. *Annalen der Physik*, **17**, 132–84.

Elsasser, W. M. 1942: *Heat Transfer by Infrared Radiation in the Atmosphere*. Cambridge, MA: Harvard University, Harvard Meteorological Studies no. 6.

Erickson, G. J. and Smith, C. R. (eds) 1988: *Maximum Entropy and Bayesian Methods in Science and Engineering*. Boston, MA: Kluwer Academic Publishers.

Evans, J. W. 1965: The sun. In Valley (1965).

Fahrenheit, M. 1724: Experiments concerning the degrees of heating of boiling liquors. *Philosophical Transactions of the Royal Society of London*, **7**, 1–2. [Abridged version of *Transactions*.]

Feigenbaum, L. 1981: *Brook Taylor's Methodus Incrementorum: A Translation with Mathematical and Historical Commentary*. PhD thesis, Yale University.

Fisher, R. A. 1920: A mathematical examination of the methods of determining the accuracy of an observation by the mean error, and by the mean square error. *Monthly Notices of the Royal Astronomical Society*, **LXXX**, 758–70. [Also printed in Fisher (1950) as paper #2.]

Fisher, R. A. 1950: *Contributions to Mathematical Statistics*. New York: John Wiley & Sons, Inc.

Fleming, J. R. (ed.) 1996: *Historical Essays on Meteorology 1919–1995*. Boston, MA: The American Meteorological Society.

Ford, K. W. 1968: *Basic Physics*. Waltham, MA: Ginn-Baisdell.

Fourier, J. B. J. 1822: *Theorie analytique de la chaleur*. Dover. [Translated by A. Freeman.]

Fourier, J. B. J. 1824: Remarques général sur les températures du globe terrestre et des espaces planétaires. *Annales de chimie et de physique*, **27**, 136–67.

Frankland, E. 1861: On the physical cause of the glacial epoch. *Philosophical Magazine*, **Ser4 vol. 27**, 321–41.

Fresnel, A. J. 1818: *Mémoire sur la diffraction de la lumière*. [Memoir on the diffraction of light. Also see Landmarks of Science, Readex Microprint Corp., NY, 1970.]

Friedman, R. M. 1989: *Appropriating the weather: Vilhelm Bjerknes and the Construction of Modern Meteorology*. Ithaca, NY: Cornell University Press.

Fultz, D. 1951: Experimental analogies to atmospheric motions. In Malone (1951), 1235–48.

Galton, F. 1883: *Inquiries into Human Faculty and its Development*. London: Macmillan.

Gast, P. R. 1965: Solar Electromagnetic Radiation, 16.1 Solar irradiance 16.1–16.9. In Valley (1965).

Gedzelman, S. D. 1994: Chaos rules. *Weatherwise* 21–6.

Gentleman, W. M. and Sande, G. 1966: Fast Fourier transforms – for fun and profit. In *1966 Fall Joint Computer Conference, AFIPS Proc.*, Washington, DC: Spartan, 563–78.

Gibbs, J. W. 1872: Graphical methods in the thermodynamics of fluids. *Transactions of the Connecticut Academy of Arts and Sciences*, **II**, 309–42.

Gillispie, C. C. (ed.) 1981: *Dictionary of Scientific Biography*. New York: Charles Schribner's Sons. [Several volumes.]

Gleick, J. 1987: *Chaos: making a new science?* New York: Penguin.

Godson, W. L. 1950: A study of the deviations of wind speeds and directions from geostrophic values. *Quarterly Journal of the Royal Meteorological Society*, **76**, 3–15.

Goff, J. A. and Gratch, S. 1946: Low-pressure properties of water from -160 to 212 F. *Heating. Piping and Air Conditioning*, **18**, 125–36. [American Society of Heating and Ventilating Engineers, Journal Section, February 1946. Presented at the 52nd Annual Meeting in New York, January 1946. Also see note March 1946, p. 125.]

Goossens, M., Mittelbach, F., and Samarin, A. 1994: *The LATEX Companion*. Reading, MA: Addison-Wesley.

Grattan-Guinness, I. 1981: Pierre-Simon, Marquis de Laplace. In Gillispie (1981).

Gray, W. M. 1978: Hurricanes: their formation, structure and likely role in the tropical circulation. In Shaw (1978).

Grmek, M. D. 1981: Santorio Santorio. In Gillispie (1981).

Guerlac, H. 1981: Antoine-Laurent Lavoisier. In Gillispie (1981).

Gurney, R. J., Foster, J. L., and Parkinson, C. L. (eds) 1993: *Atlas of Satellite Observations Related to Global Change*. Cambridge, England: Cambridge University Press.

Hadley, G. 1735: On the cause of the general trade-winds. *Philosophical Transactions of the Royal Society of London*, **29**, 58–62.

Haines, D. A. and Winston, J. S. 1963: Monthly mean values and spatial distribution of meridional transport of sensible heat. *Monthly Weather Review*, **91**, 319–28.

Haltiner, G. J. and Martin, F. L. 1957: *Dynamical and Physical Meteorology*. New York: McGraw-Hill Book Co.

Haltiner, G. J. and Williams, R. T. 1980: *Numerical Prediction and Dynamic Meteorology*. New York: John Wiley & Sons, Inc.

Harington, C. R. (ed.) 1992: *The Year Without a Summer? World Climate in 1816*. Ottawa: Canadian Museum of Nature.

Harris, B. (ed.) 1967: *Spectral Analysis of Time Series*. New York: John Wiley & Sons, Inc. Proceedings of an Advanced Seminar Conducted by the Mathematics Research Center, United States Army and The Statistics Department at the University of Wisconsin, Madison October 3–5, 1966.

Hasselmann, K., Munk, W., and MacDonald, G. 1963: Bispectra of ocean waves. In Rosenblatt (1963), 125–39.

Haurwitz, B. 1941: *Dynamic Meteorology*. New York: McGraw-Hill Book Co.

Hawkins, G. S. 1965: *Stonehenge Decoded*. Garden City, NY: Doubleday & Company, Inc.

Hide, R. 1953: Some experiments on thermal convection in a rotating liquid. *Quarterly Journal of the Royal Meteorological Society*, **79**, 161.

Hide, R. 1983: Dynamics of rotating fluids and planetary atmospheres. *Quarterly Journal of the Royal Meteorological Society*, **Special**, 37–45. Recent advances in meteorology and physical oceanography. Reviews presented at a meeting held in London in October 1983 to celebrate the centenary of the granting of the title "Royal Meteorological Society" by Queen Victoria in September 1883.

Hide, R. 1989: Sir Harold Jeffreys, F.R.S. *Quarterly Journal of the Royal Meteorological Society*, **115**, 711–13.

Hiebert, E. H. 1962: *Historical Roots of the Principle of Conservation of Energy*. The State Historical Society of Wisconsin. University Microfilms, Ann Arbor, 1967.

Holloway, J. L. 1958: Smoothing and filtering of time and space series. *Advances in Geophysics*, **4**, 351–89. [New York, NY: Academic Press.]

Holper, P. 1999: Personal communication. Communication Manager, CSIRO Atmospheric Research, Australia.

Holton, J. R. 1979: *An introduction to dynamic meteorology*. New York: Academic Press.

Houghton, H. G. 1954: On the heat balance of the northern hemisphere. *Journal of Meteorology*, **11**, 1–9.

Howard, J. N. 1965: Transmission and detection of infrared radiation, 10.2 Atmospheric absorption 10.2–10.6. In Valley (1965).

Howarth, D. A. and Rayner, J. N. 1991: An analysis of the water vapor flux divergence field over the southern hemisphere. *Physical Geography*, **14**, 513–35.

Hutchings, J. W. 1957: Vapor flux. *Quarterly Journal of the Royal Meteorological Society*, **83**, 30–48.

Huygens, C. 1690: *Traité de la lumière*. [Treatise on light. Republished in English by Dover, NY, 1962.]

Imbrie, J. and Imbrie, J. Z. 1980: Modeling the climatic response to orbital variations. *Science*, **207**, 943–53.

Iribarne, J. and Godson, W. 1973: *Atmospheric Thermodynamics*. Dordrecht, Holland: D. Reidel Publishing Company.

Jayawardene, S. A. 1981: Rafael Bombelli. In Gillispie (1981).

Jeffreys, H. 1926: On the dynamics of geostrophic winds. *Quarterly Journal of the Royal Meteorological Society*, **52**, 85–104.

Jeffreys, H. 1931: *Cartesian Tensors*. Cambridge: Cambridge University Press.

Jenkins, G. M. and Watts, D. G. 1968: *Spectral Analysis and its Applications*. San Francisco: Holden-Day.

Jennison, R. C. 1961: *Fourier Transforms and Convolutions for the Experimentalist*. New York: Pergamon Press.

Joule, J. P. 1847a: On matter, living force, and heat. [In Joule (1884), 265–76 from A lecture at St. Ann's Church Reading-Room; and published in the *Manchester Courier* newspaper, May 5 and 12, 1847.]

Joule, J. P. 1847b: On the mechanical equivalent of heat, as determined from the heat evolved by the agitation of liquids. [In Joule (1884), 276–7. Rep. Brit. Assoc. 1847, Sections, p. 55. Read before the British Association at Oxford, June 1847.]

Joule, J. P. 1884: *The Scientific Papers of James Prescott Joule*. Two volumes. London: Taylor and Francis Ltd.

Kangro, H. 1981a: Max Karl Ernst Ludwig Planck. In Gillispie (1981).

Kangro, H. 1981b: Wilhelm Carl Werner Otto Fritz Franz Wien. In Gillispie (1981).

Kay, S. M. 1988: *Modern Spectral Analysis*. Englewood Cliffs, NJ: Prentice Hall.

Kendall, M. G. 1956: Studies in the history of probability and statistics. *Biometrika*, **43**, 1–11.

Kendall, M. G. and Buckland, W. R. 1971: *A Dictionary of Statistical Terms*. New York: Hafner Publishing McGraw-Hill Book Co.

Kendall, M. G. and Stuart, A. 1969: *The Advanced Theory of Statistics*. New York: Hafner Publishing McGraw-Hill Book Co. In three volumes.

Kestenbaum, D. 1998: Gravity measurements close in on Big G. *Science*, **282**, 2180–81.

Kirchhoff, G. R. 1859: Emission and absorption. *Monatsbericht der Academie der Wissenschaften zu Berlin*, **Dec.** [See translation in Magie (1963) p. 356. Title possibly not correct.]

Knuth, D. E. 1986: *The TEXbook*. Reading, MA: Addison-Wesley Publishing Company.

Korn, G. A. and Korn, T. M. 1967: *Manual of Mathematics*. New York: McGraw-Hill Book Co.

Kunde, V. G., Conrath, B. J., Hanel, R. A., Maguire, W. C., and Prabhakara, C. 1974: The Nimbus 4 infrared spectroscopy experiment. 2. Comparison of observed and theoretical radiances from 425–1450 cm^{-1}. *Journal of Geophysical Research*, **79**, 777–84. Additional author V. V. Salomonson.

Lagrange, J. L. 1811: *Mécanique analytique*. Paris: Mme Ve Courcier. [See Boissonnade and Vagliente (1997).]

Lamb, H. 1932: *Hydrodynamics*. New York: Dover. [First edition published by Cambridge University Press in 1879, Sixth edition of 1932 republished in 1945 by Dover.]

Lamb, H. 1997: *Through All the Scenes of Life: A Meteorologist's Tale*. East Harling, Norfolk, England: Taverner Publications.

Lambert, J. H. 1760: *Photometria, sive De Menura et gradibus luminis, colorum et umbrae*. V. E. Klett. [Photometry, or On the measurement and grading of light, color, and shade. Also see Landmarks of Science, Readex Microprint Corp., NY, 1970.]

Lamport, L. 1994: $\mathit{E\!T\!E\!X}$. Reading, MA: Addison-Wesley.

Landsberg, H. 1960: *Physical Climatology*. duBois, PA: Gray Printing Company.

Landsberg, H. E. and Jacobs, W. C. 1951: Applied Climatology. In Malone (1951), 976–92.

Landsberg, H. E. and van Miegham, J. (eds) 1964: *Advances in Geophysics*. Boston, MA: Academic Press. [vol. 10. Series has had several different editors.]

Laplace, P. S. 1825: *Traité de Méchanique Célestre*. Paris. [Nathaniel Bowditch translated the first four volumes of Marquis de Laplace's Celestial Mechanics into English and included extensive comments. Volume 5 remains in French. The French originals have dates 1799–1825. The Bowditch versions were last published by the Chelsea Pub. Co., Bronx, NY in 1966–69.]

Lavoisier, A. L. 1789: *Traité élémentaire de chimie*. Paris: Chez Chuchet. [Translated section contained in Lindsay (1975).]

Lean, J. and Rind, D. 1998: Climate forcing by changing solar radiation. *Journal of Climate*, **11**, 3069–94.

Lindsay, R. (ed.) 1975: *Energy: Historical Development of the Concept*. Stroudsburg, PA: Dowden, Hutchinson & Ross, Inc.

List, R. J. (ed.) 1958: *Smithsonian Meteorological Tables*. Washington, DC: The Smithsonian Institution. Smithsonian Miscellaneous Collections, vol. 114.

Liu, K. and Shen, C. 1998: Reconstruction of a 1,000 year record of typhoon landfalls in southern China from Chinese historical documentary sources. Workshop on calibration of historical data for reconstruction of climate variations. Barcelona, Spain, 6–8 July, National Oceanographic and Atmospheric Administration and National Science Foundation.

Lorenz, E. N. 1955: Available potential energy and the maintenance of the general circulation. *Tellus*, **VII**, 157–67.

Lorenz, E. N. 1967: *The Nature of the General Circulation of the Earth*. Geneva, Switzerland: World Meteorological Organization, (WMO No. 218.TP.115).

Lorenz, E. N. 1968: Climatic determinism. *Meteorological Monographs*, **8**, 1–3.

Lorenz, E. N. 1975: Climate predictability. In WMO (1975) Appendix 2.1, 132–6.

Lorenz, E. N. 1976: Nondeterministic theories of climatic change. *Quaternary Research*, **6**, 495–506.

Lorenz, E. N. 1977: Victor Paul Starr. *Quarterly Journal of the Royal Meteorological Society*, **103**, 222–3.

Lorenz, E. N. 1993: *The Essence of Chaos*. Seattle, WA: University of Washington.

Lorenz, E. N. 1996: The evolution of dynamic meteorology. In Fleming (1996) 1–19.

Lyell, C. 1873: *Principles of Geology*. New York, Appleton.

Magie, W. F. 1963: *A Source Book in Physics*. Cambridge, MA: Harvard University Press. [First Edition 1935.]

Malone, T. (ed.) 1951: *Compendium of Meteorology*. Boston, MA: American Meteorological Society.

Manabe, S. (ed.) 1985: *Issues in Atmospheric and Oceanic Modeling*. New York: Academic Press. Advances in Geophysics V28.

Manabe, S. and Broccoli A. J. 1990: Mountains and arid climates of middle latitudes. *Science*, **247**, 192–5.

Maor, E. 1994: *e the story of a number*. Princeton, NJ: Princeton University Press.

Margules, M. 1901: The mechanical equivalent of any given distribution of atmospheric pressure, and the maintenance of a given difference in pressure. *Smithsonian Miscellaneous Collections*, **51**, 501–32. [Translated by Cleveland Abbe, 1910.]

Margules, M. 1905: The energy of storms. *Smithsonian Miscellaneous Collections*, **51**, 533–95. [Translated by Cleveland Abbe, 1910.]

Maxwell, J. C. 1865: A dynamical theory of the electromagnetic field. *Philosophical Transactions of the Royal Society of London*, **155**, 459–512. [Also see book with same title, Scottish Academic Press, 1982.]

Maxwell, J. C. 1881: *A Treatise on Electricity and Magnetism*. Oxford: Clarendon Press. In two volumes.

Mayow, A. J. 1668: An account of two books. *Philosophical Transactions of the Royal Society of London*, **III**, 295–8.

McGuffie, K. and Henderson-Sellers A. 1997: *A Climate Modelling Primer*. New York: John Wiley & Sons, Inc.

McKeon, R. M. 1981: Claude-Louis-Marie-Henri Navier. In Gillispie (1981).

Meeus, J. 1991: *Astronomical Algorithms*. Richmond, VA: Willmann-Bell.

Mendoza, E. (ed.) 1977: *Reflections on the Motive Power of fire by Sadi Carnot and Other Papers on the Second Law of Thermodynamics by E. Clapeyron and R. Clausius*. Gloucester, MA: Peter Smith. [Copyright 1960 by Dover Publications, Inc.]

Mesinger, F. 1984: A blocking technique for representation of mountains in atmospheric models. *Rivista di Meteorologia Aeronautica*, **44**, 195–202.

Middleton, W. 1969: *The Invention of Meteorological Instruments*. Baltimore, MD: The Johns Hopkins Press.

Middleton, W. E. K. (ed.) 1961: *Optical Treatise on the Gradation of Light*. Canada: University of Toronto Press.

Mie, G. 1908: Beitrage zur Optik trüber Medien, speziell kolloidaler Metallösungen. *Ann. Physik*, **25**, 377–445.

Milankovitch, M. 1969: *Canon of Insolation and the Ice-age Problem*. Israel Program for Scientific Translations, Jerusalem. [Original title: Kanon der Erdbestrahlung und seine Anwendung auf das Eiszeitenproblem, Royal Serbian Academy Special Publications, vol. 132, Section of Mathematical and Natural Sciences, vol. 33, 1941.]

Millar, W. J. (ed.) 1881: *Miscellaneous scientific papers: by W. J. M. Rankine*. London: Charles Griffin and Company.

Miller, D. H. 1968: A survey course: The energy and mass budget at the earth's surface. *Association of American Geographers, Commission on College Geography, Publication*, **7**.

Mintz, Y. 1951: The geostrophic poleward flux of angular momentum in the month of January 1949. *Journal of Meteorology*, **3**, 15–20.

Mintz, Y. and Dean, G. 1952: The observed mean field of motion of the atmosphere. *Geophysical Research Papers*, **17**. Cambridge, MA: Air Force Cambridge Research Center.

Moellering, H. and Rayner, J. N. 1981: The harmonic analysis of spatial shapes using dual axis Fourier spae analysis (DAFSA). *Geographical Analysis*, **13**, 64–77.

Mood, A. M. and Graybill, F. A. 1963: *Introduction to Statistics*. New York: McGraw-Hill Book Co.

More, L. T. 1934: *Isaac Newton, A Biography 1642–1727*. New York: Dover Publications Inc.

Namias, J. 1983: The history of polar front and air mass concepts in the United States – An eyewitness account. *Bulletin of the American Meteorological Society*, **64**, 734–55.

Navier, C. L. M. H. 1822: Mémoire sur les lois du mouvement des fluides. *Mémoires de l'Academie des Sciences de Berlin*, **vi**, 389.

Newell, R. E., Vincent D. G., Doplick T. G., Ferruzza D., and Kidson J. W., 1969: The energy balance of the global atmosphere. In Corby (1969) 42–90.

Newton, I. 1687: *Philosophiæ Naturalis Principia Mathematica*. Berkeley: University of California Press (1966). [Motte's Translation Revised by Florian Cajori in two volumes.]

Newton, I. 1701: A scale of the degrees of heat. *Philosophical Transactions of the Royal Society of London*, **4**, 572–5. [Abridged version of *Transactions*.]

Newton, I. 1704: *Opticks*. London.

Norske-Videnskaps (ed.) 1962: *In Memory of Vilhelm Bjerknes on the 100th Anniverasry of his birth*. Norske videnskaps – akademi i Oslo. Geophysiske publikasjoner, Bd 24. In cooperation with the American Meteorological Society, Oslo, Norway.

Oxford (ed.) 1978: *The Oxford English Dictionary*. Oxford: Clarendon Press.

Parkinson, C. L. 1985: *Breakthroughs: A Chronology of great Achievements in Science and Mathematics*. Boston, MA: G. K. Hall and Co.

Parkinson, E. M. 1981: George Gabriel Stokes. In Gillispie (1981).

Pearson, K. 1894: Contributions to the mathematical theory of evolution. *Philosophical Transactions of the Royal Society of London*, **185A**, 71–110.

Peixoto, J. P. and Oort A. H. 1992: *Physics of Climate*. New York: American Institute of Physics.

Peixoto, J. P., Saltzman B., and Teweles, S. 1964: Harmonic analysis of the topography along parallels of the earth. *Journal of Geophysical Research*, **69**, 1501–5.

Persson, A. 1998: How do we understand the Coriolis force? *Bulletin of the American Meteorological Society*, **79**, 1373–85.

Phillips, N. A. 1957: A coordinate system having some special advantages for numerical forecasting. *Journal of Meteorology*, **14**, 184–5.

Phillips, N. A. 1982: Jule Charney's influence on meteorology. *Bulletin of the American Meteorological Society*, **63**, 492–8.

Phillips, O. M. 1968: *The Heart of the Earth*. San Francisco, CA: Freeman.

Platzman, G. W. 1987: Conversations with Jule Charney. Technical Report NCAR Technical Note 298, Boulder,CO: National Center for Atmospheric Research.

Poisson, S. D. 1829: Mémoire sur les equations générales de l'équilibre et du mouvement des corps solides élastique et des fluides. *Journal de l'École Polytechnique*, **xiii**, 1.

Priestley, C. H. B. 1949: Heat transport and zonal stress between latitudes. *Quarterly Journal of the Royal Meteorological Society*, **75**, 28–40.

Priestley, C. H. B. and Swinbank W. C. 1947: Vertical transport of heat by turbulence in the atmosphere. *The Proceedings of the Royal Society of London*, **A189**, 543–61.

Ptolemaeus, C. 1984: *Almagest*. London, England: Duckworth. [The title is: Ptolemy's Almagest translated and annotated by G. J. Toomer.]

Queney, P. 1948: The problem of air flow over mountains: a summary of theoretical studies. *Bulletin of the American Meteorological Society*, **29**, 16–26.

Ramsay, W. 1896: *The Gases of the Atmosphere, The History of Their Discovery*. New York: Macmillan.

Rankine, W. J. M. 1855: Outlines of the science of energetics. *Proceedings of the Philosophical Society of Glasgow*. [Published in Millar (1881), 209–28.]

Rayner, J. N. 1971: *An Introduction to Spectral Analysis*. London: Pion Limited.

Rayner, J. N. 1973a: Scale analysis of skewness: the bispectrum. In *Proceedings of the Association of Geographers*, **5**, Washington, DC, 230–3.

Rayner, J. N. 1973b: The practical application of one dimensional spectral analysis. *Geographia Polonica*, **25**, 67–92.

Rayner, J. N., Hobgood J. S., and Howarth D. A. 1991: Dynamic climatology: its history and future. *Physical Geography*, **12** (3), 207–19.

Reiter, E. R. 1969: *Atmospheric Transport Processes*. Oak Ridge, TN: US Atomic Energy Commission. [In three volumes (TID-24868, 25314, and 25731) 1969, 1971, and 1972, available from the Clearinghouse for Federal Scientific and Technical Information, National Bureau of Standards, Va 22151.]

Reynolds, O. 1883: An experimental investigation of the circumstances which determine whether the motion water shall be direct or sinuous, and of the law of resistance in parallel channels. *Philosophical Transactions of the Royal Society of London*, **Part III**, 935–82.

Reynolds, O. 1894: On the dynamical theory of incompressible viscous fluids and the determination of the criterion. *Philosophical Transactions of the Royal Society of London*, **136**, 123–64.

Richardson, L. F. 1910: The approximate arithmetical solution by finite differences of physical problems involving differential equations, with an application to the stresses in a masonry dam. *Philosophical Transactions of the Royal Society of London*, **210**, 307–57.

Richardson, L. F. 1920: The supply of energy from and to atmospheric eddies. *The Proceedings of the Royal Society of London*, **A97**, 354–73.

Richardson, L. F. 1922: *Weather Prediction by Numerical Process*. Cambridge, UK: Cambridge University Press. [Dover publication 1965.]

Richtmyer, R. D. and Morton, K. W. 1967: *Difference Methods for Initial-value Problems*. New York: John Wiley & Sons.

Robbins, A. 1994: What's GNU? *Linux Journal*, **1**, 20–5.

Rodhe, H., Charleson, R. , and Crawford, E. 1997: Svante Arrhenius and the greenhouse effect. *Ambio: A Journal of the Human Environment*, **XXVI**, 2–5.

Roller, D. 1950: *The Early Development of the Concepts of Temperature and Heat*. Cambridge, MA: Harvard University Press. Harvard Case Histories in Experimental Science, Case 3.

Rosenblatt, M. (ed.) 1963: *Proceedings of Symposium on Time Series*. New York: John Wiley & Sons, Inc.

Rosenfeld, L. 1981: Gustav Robert Kirchhoff. In Gillispie (1981).

Rossby, C. G. 1941: *The Scientific Basis of Modern Meteorology*. Washington, DC: US Government Printing Office. [Republished in Berry et al. (1945) Section VII, 502–29.]

Rossby, C. G. and Collaborators 1939: Relation between variations in the intensity of the zonal circulation of the atmosphere. *Journal of Marine Research*, **2**, 38.

Roth, L. 1937: *Descartes' Discourse on Method*. Oxford: Claredon Press.

Rumford, C. 1798: An experimental inquiry concerning the source of the heat which is excited by friction. *Philosophical Transactions of the Royal Society of London*, **LXXXVIII**, 80–102.

Safire, W. 1990: Greenhouse effect. *New York Times*. Sunday March 5.

Saltzman, B. 1957: Equations governing the energetics of the larger scales of atmospheric turbulence in the domain of wave number. *Journal of Meteorology*, **14**, 513–23.

Saltzman, B. (ed.) 1962: *Selected Papers on the Theory of Thermal Convection with Special Application to the Earth's Planetary Atmosphere*. New York: Dover.

Saltzman, B. and Fleisher, A. 1960a: The exchange of kinetic energy between the larger scales of atmospheric motion. *Tellus*, **XII**, 374–7.

Saltzman, B. and Fleisher, A. 1960b: The modes of release of available potential energy in the atmosphere. *Journal of Geophysical Research*, **65**, 1215–22.

Saltzman, B. and Teweles S. 1964: Further statistics on the exchange of kinetic energy between harmonic components of the atmospheric flow. *Tellus*, **XVI**, 432–5.

Schmidt, W. 1925: *Der Massenaustausch in freier Luft und verwandte Erscheinungen*. Hamburg, Germany: H. Grand. [The bulk exchange coefficient in the free air and associated phenomena.]

Scorer, R. S. 1949: Theory of waves in the lee of mountains. *Quarterly Journal of the Royal Meteorological Society*, **75**, 41–56.

Scorer, R. S. and Wilkinson, M. 1956: Waves in the lee of an isolated hill. *Quarterly Journal of the Royal Meteorological Society*, **82**, 419–27.

Scriba, C. J. 1981: Johann Heinrich Lambert. In Gillispie (1981).

Sears, F. W. 1953: *An Introduction to Thermodynamics, The Kinetic Theory of Gases, and Statistical Mechanics*. Reading, MA: Addison-Wesley.

Selby, S. M. (ed.) 1967: *Standard Mathematical Tables*. Cleveland, Ohio: The Chemical Rubber Company. [Various years.]

Sellers, W. D. 1965: *Physical Climatology*. Chicago, IL: University of Chicago Press.

Shapiro, A. H. 1961: *Shape and Flow: The Fluid Dynamics of Drag*. Garden City, NY: Anchor Books, Doubleday & Company, Inc.

Shapiro, A. H. 1972: *Illustrated Experiments in Fluid Mechanics*. Cambridge, MA: The MIT Press. [Based on material in films produced under direction of The National Committee of Fluid Mechanics Films. Films available from Encyclopaedia Britannica Educational Corp., Chicago, IL.]

Shaw, D. B. (ed.) 1978: *Meteorology Over the Tropical Oceans*. Bracknell, England: Royal Meteorological Society. Joint conference of the Royal Meteorological Society, the American Meteorological Society, the Deutsche Meteorologische Gesellschaft and the Royal Society.

Sheppard, W. F. 1899: Central-difference formulae. *Proceedings London Mathematical Society*, **xxxi**, 449–88.

Simpson, J. E. 1994: *Sea Breeze and Local Winds*. Cambridge, England: Cambridge University Press.

Smith, D. E. 1923: *History of Mathematics*. New York: Dover Publications, Inc. [Dover: 1951–53.]

Smith, M. L. and Dahlen, F. A. 1981: The period and Q of the Chandler wobble. *Geophysical Journal of the Royal Astronomical Society*, **64**, 223–81.

Stallman, R. 1997: *Gnu emacs manual*. Boston, MA: Free Software Foundation.

Starr, V. P. 1948: An essay on the general circulation of the earth's atmosphere. *Journal of Meteorology*, **5**, 39–43.

Starr, V. P. 1968: *Physics of Negative Viscosity Phenomena*. New York: McGraw-Hill Book Co.

Starr, V. P. and White, R. M. 1952: Schemes for the study of hemispheric exchange processes. *Quarterly Journal of the Royal Meteorological Society*, **78**, 407–10.

Starr, V. P. and White, R. M. 1954: Balance requirements of the general circulation. *Geophysical Research Papers*, **35**. Cambridge, MA: Air Force Cambridge Research Center.

Stefan, J. 1879: Über die Beziehung zwischen der Wärmaestrahlung und der Temperatur. *Sitzungsberichte der Akademie der Wissenschaften in Wien*, **79, ptII**, 391. [See Magie (1963), 377 for translation.]

Stokes, S. G. G. 1845: On the theories of the internal friction of fluids in motion. *Cambridge Philosophical Transactions*, **x**, 287–319.

Straub, H. 1981: Daniel Bernoulli. In Gillispie (1981).

Stuewer, R. H. 1981: Jean Baptiste Perrin. In Gillispie (1981).

Sutcliffe, R. C. 1958: Professor C.-G Rossby. *Quarterly Journal of the Royal Meteorological Society*, **84**, 88–9.

Sutton, O. G. 1953: *Micrometeorology*. New York: McGraw-Hill Book Co.

Taylor, G. I. 1915: Eddy motion in the atmosphere. *Philosophical Transactions of the Royal Society of London*, **215**, 1–26.

Thackray, A. 1981: John Dalton. In Gillispie (1981).

Thompson, P. D. 1961: *Numerical Weather Analysis and Prediction*. New York: Macmillan.

Thompson, P. D. 1983: A history of numerical weather prediction in the United States. *Bulletin of the American Meteorological Society*, **64**, 755–69.

Trenberth, K. E. (ed.) 1992: *Climate System Modeling*. Cambridge, England: Cambridge University Press.

Trewartha, G. T. 1954: *An Introduction to Climate*. New York: McGraw-Hill Book Co. [Only the third edition contains the material referenced in the text.]

Tukey, J. W. 1967: An introduction to the calculations of numerical spectrum analysis. In Harris (1967).

Turco, R. P. 1995: *Earth under Siege*. Oxford: Oxford University Press.

Tyndall, J. 1861: On the absorption and radiation of heat by gases and vapours, and on the physical connexion of radiation, absorption, and conduction. *Philosophical Transactions*, **xxii**, 169.

Tyndall, J. 1873: *Contributions to Molecular Physics in the Domain of Radiant Heat*. New York: D. Appleton and Company.

Ulam, S. 1958: John von Neumann. *Bulletin of the American Mathematical Society*, **64**, 1–49. [The May issue was devoted to von Neumann.]

Valley, S. L. (ed.) 1965: *Handbook of Geophysics and Space Environments*. New York: McGraw-Hill Book Co.

Vernekar, A. D. 1972: Long-term global variations of incoming solar radiation. *Meteorological Monographs*, **12, No. 34**.

Vinnichenko, N. K. 1970: The kinetic energy spectrum in the free atmosphere – 1 second to 5 years. *Tellus*, **22**, 158–66.

Volkerding, P., Reichard, K., and Johnson, E. F. 1996: *Linux*. New York, 10011, MIS: Press, Henry Holt and Co., Inc.

von Helmholtz, H. L. F. 1858: On the integrals of the hydrodynamic equations that represent vortex-motions. *Crelle's Journal für die reine und angewandte Mathematik*, **LV**, 25–85. (see Abbe, Cleveland, 1891: The Mechanics of the Earth's Atmosphere. A Collection of Translations. Smithsonian miscellaneous collections, 31–66.)

von Mayer, J. R. 1842: Bemerkungen über die Kräfte der unbelebten Natur. *Wöhler and Liebig's Annalen der Chemie und Pharmacie*, **xlii**, 239. [A full translation of von Mayer's paper, "On the forces of inorganic nature," and heat conversion magnitude is given in Lindsay (1975).]

Washington, W. M. and Parkinson, C. L. 1986: *An Introduction to Three-dimensional Climate Modeling*. Mill Valley: University Science Books.

Weast, R. C. (ed.) 1965: *Handbook of Chemistry and Physics*. Cleveland, OH: The Chemical Rubber Co. [This reference is published annually.]

Welsh, M. and Kaufman, L. 1995: *Running LINUX*. Sebastopol, CA: O'Reilly & Associates, Inc.

Whiteside, D. T. (ed.) 1964: *The Mathematical Works of Isaac Newton*. New York: Johnson Reprint Corporation. [Contains English translations of Newton's work including the 1737 "A treatise of the method of fluxions and infinite series, with its application to the geometry of curved lines," from the unpublished Latin.]

Wiin-Nielsen, A., Brown, J. A., and Drake, M. 1963: On atmospheric energy conversions between the zonal flow and the eddies. *Tellus*, **XV**, 261–79.

Willett, H. C. 1931: Ground plan for dynamic meteorology. *Monthly Weather Review*, **59**, 219–23. [See p. iv of the index volume correcting the title to read "climatology" not "meteorology."]

Williams, J., Barry, R. G., and Washington, W. M. 1974: Simulation of the atmospheric circulation using the NCAR global circulation model. *Journal of Applied Meteorology*, **13**, 305–17.

Willson, R. C. 1993: Solar irradiance. In Gurney et al. (1993), 5–18.

Wilson, E. R. 1901: *Vector Analysis – founded on the Lectures of J. Willard Gibbs*. New Haven, CT: Yale University.

WMO, 1972: *Parameterization of Sub-grid Scale Processes*. International Council of Scientific Unions and World Meteorological Organization. GARP Publication Series No. 8.

WMO, 1973: *Methods for the Approximate Solution of Time Dependent Problems*. International Council of Scientific Unions and World Meteorological Organization. GARP Publication Series No. 10, authors H. Kreiss and J. Oliger.

WMO, 1975: *The Physical Basis of Climate and Climate Modelling*. International Council of Scientific Unions and World Meteorological Organization. WMO-ICSU Joint Organizing Committee. GARP Publication Series No. 16.

WMO, 1976: *Numerical Methods Used in Atmospheric Models*. International Council of Scientific Unions and World Meteorological Organization. WMO-ICSU Joint Organizing Committee. GARP Publication Series No. 17 in two volumes.

Yamamoto, G. 1952: On a radiation chart. *Science Report*, **5**. Tohoku University, Japan, Geophysics 4: 9–23.

Young, T. 1807: *A Course of Lectures on Natural Philosophy and the Mechanical Arts.* London: Joseph Johnson.

Youschkevitch, A. P. 1981: Leonhard Euler. In Gillispie (1981).

Zehnder, J. A. 1991: The interaction of planetary-scale tropical easterly waves with topography: a mechanism for the initiation of tropical cyclones. *Journal of the Atmospheric Sciences*, **48**, 1217–30.

Index

9 781577 180166